The World's Best Anatomical Charts

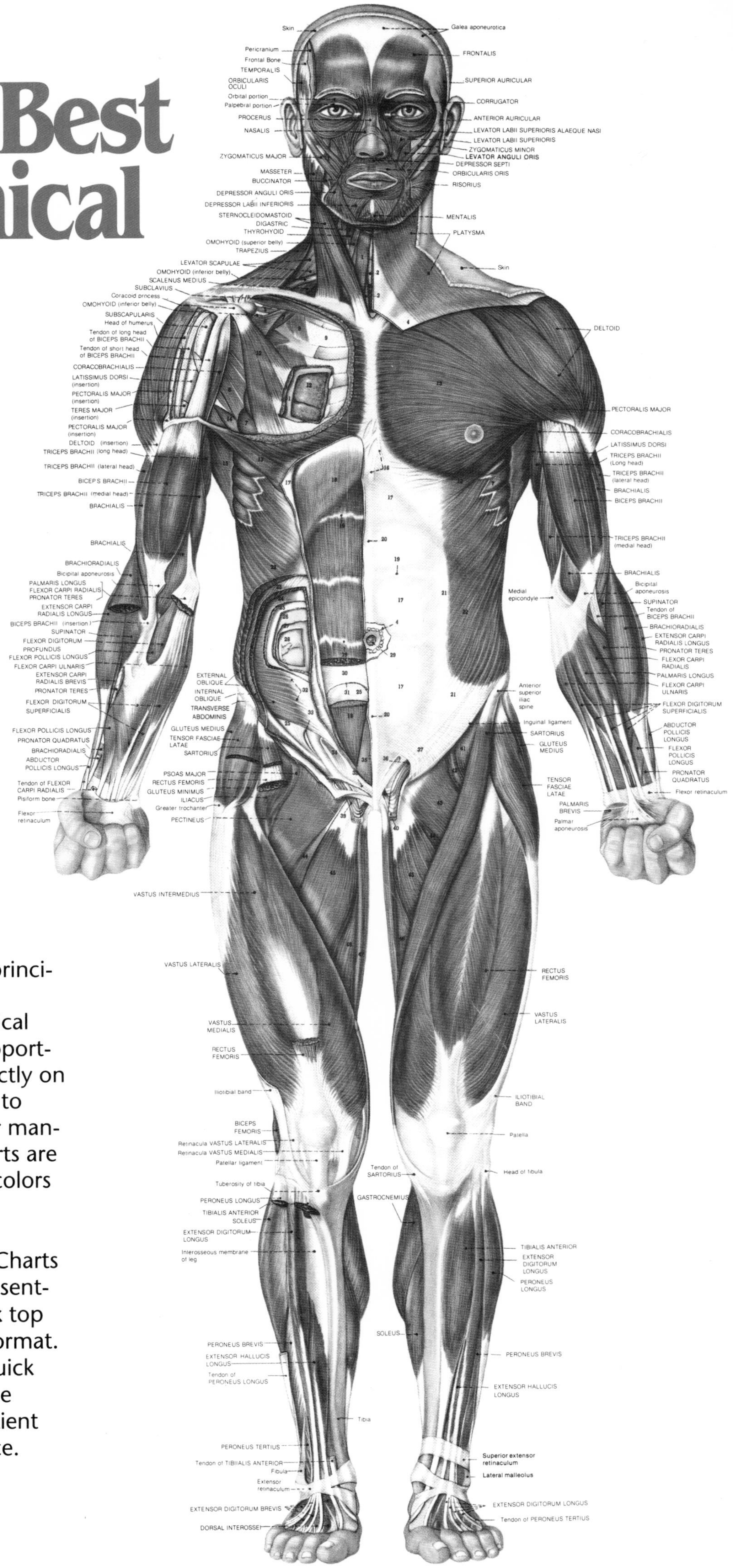

Human anatomy comes alive with our collection of 37 anatomical charts created by the world's best medical illustrators. Since 1947, the Anatomical Chart Company has been publishing educational wall charts on human anatomy that have played a vital role in medical education. Our medical illustrators work closely with physicians, researchers and scientists to create finished artwork that functions as standard reference material on a variety of important topics. Anatomical charts show the human body in a format that provides a clearer understanding of human anatomy, a visual comparison of physiological principles and highlights important pathological conditions. Medical terminology and updated supporting text is always printed directly on each chart so you never have to refer to a separate key card or manual. All of our anatomical charts are lithographed in vivid, lifelike colors on high-quality coated paper.

The World's Best Anatomical Charts features 37 full color titles presented in a handy 11" x 14" desk top size and innovative tear out format. Each chart is perforated for quick and easy removal. Ideal for the study of human anatomy, patient consultation or quick reference.

Table of Contents:

	Chart No.
Systems of the Body	
The Muscular System	1
The Skeletal System	2
The Nervous System	3
The Lymphatic System	4
The Digestive System	5
The Respiratory System	6
The Urinary Tract	7
The Female Reproductive System	8
The Male Reproductive System	9
The Endocrine System	10
The Vascular System and Viscera	11
Structures/Organs	
The Heart	12
The Brain	13
The Vertebral Column	14
The Human Skull	15
The Eye	16
The Eye: Anterior & Posterior Chambers	17
The Ear-Organs of Hearing and Balance	18
Ear, Nose and Throat	19
The Skin	20
The Human Hair	21
Shoulder and Elbow	22
Hand and Wrist	23
Hip and Knee	24
Foot and Ankle	25
The Teeth	26
Pharynx & Larynx	27
Pregnancy and Birth	28
Diseases/Disorders	
Cardiovascular Disease	29
Whiplash Injuries of the Head and Neck	30
The Human Spine-Disorders	31
Diseases of the Digestive System	32
Understanding Hypertension	33
Understanding Stroke	34
Understanding HIV & AIDS	35
Understanding Middle Ear Infection	36
Understanding the Common Cold	37

Library of Congress Catalog Card Number 93-072484
ISBN 0-9603730-5-5

THE MUSCULAR SYSTEM

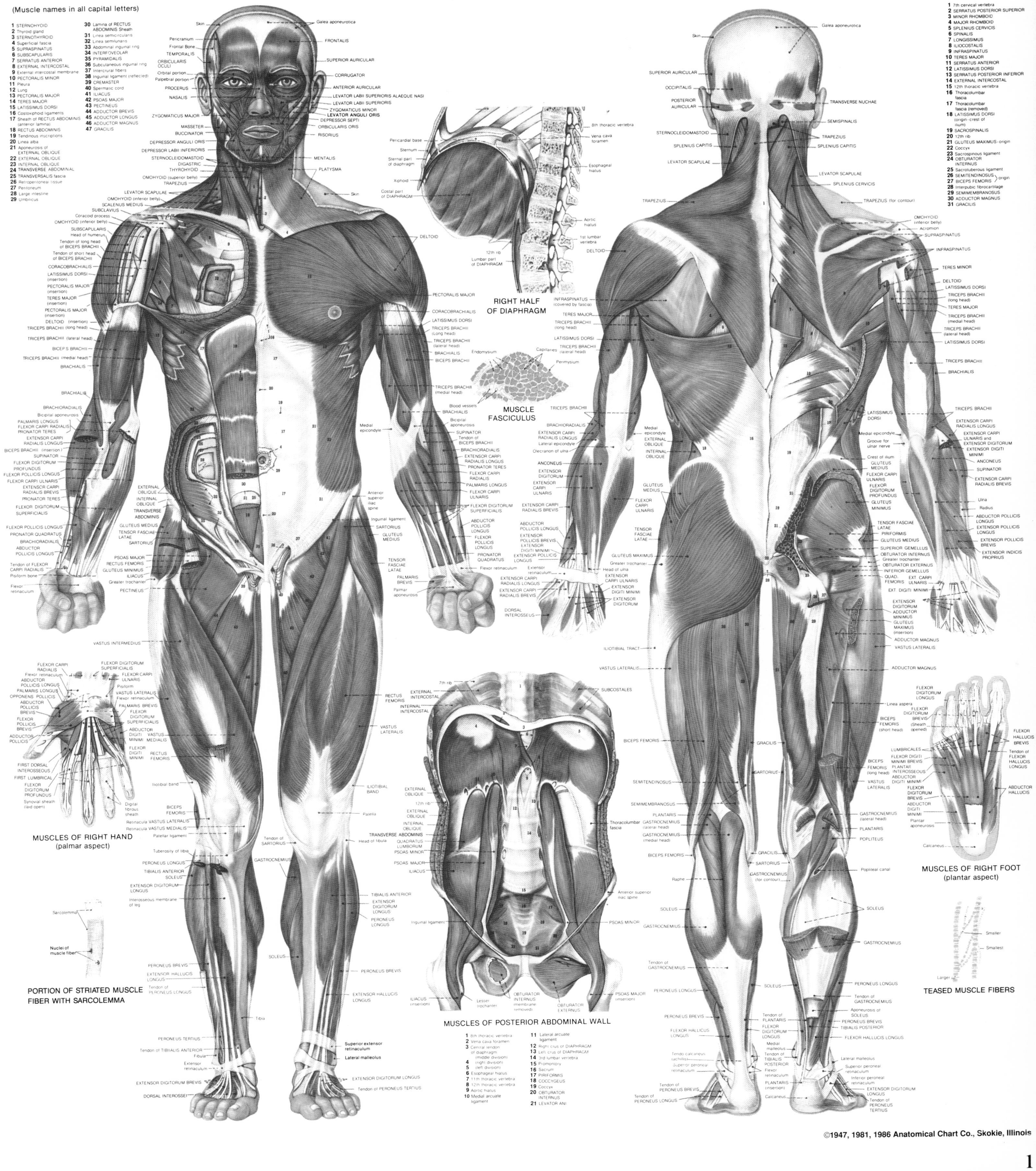

THE SKELETAL SYSTEM

(Names of bones in all capital letters)

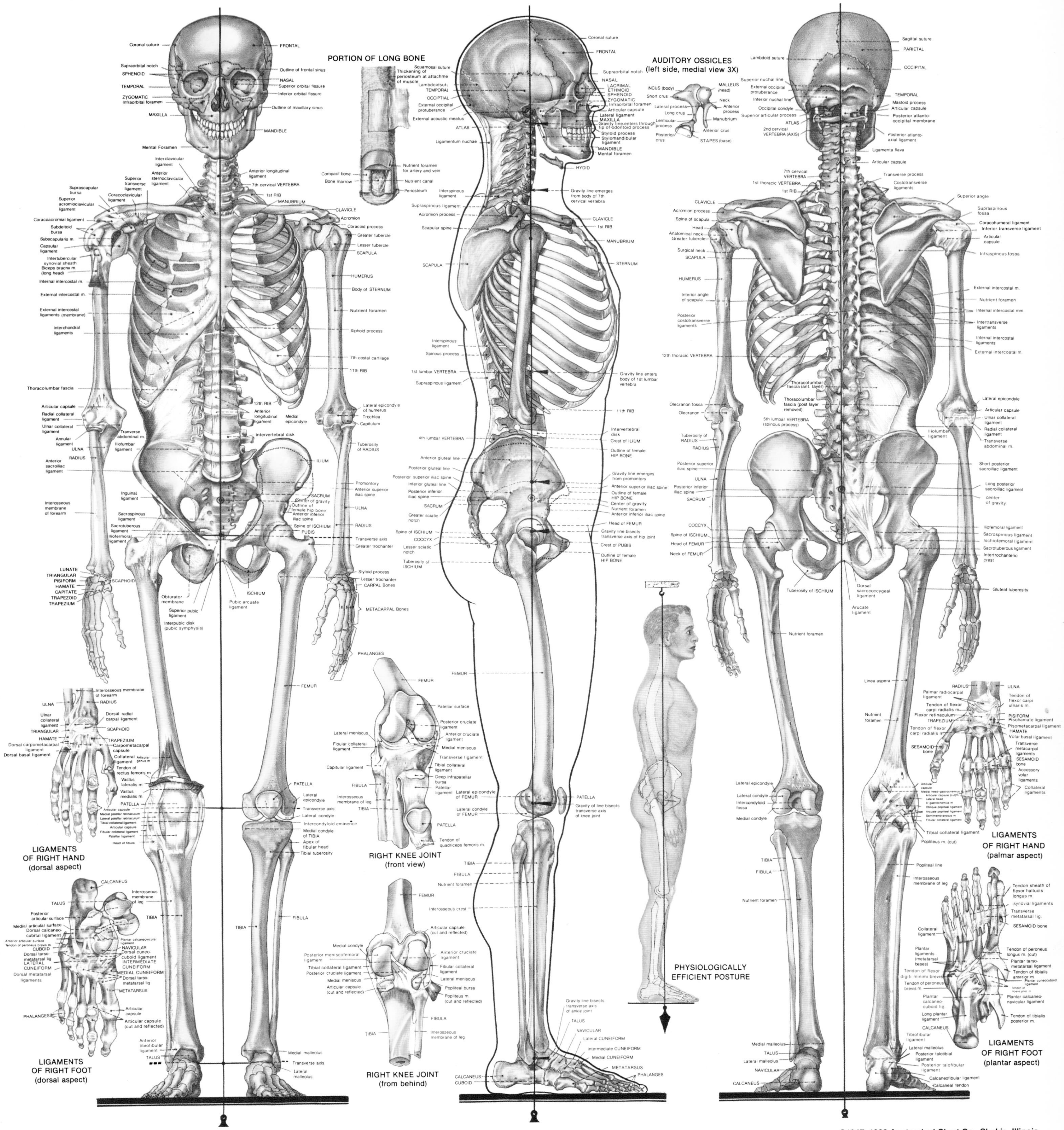

THE NERVOUS SYSTEM

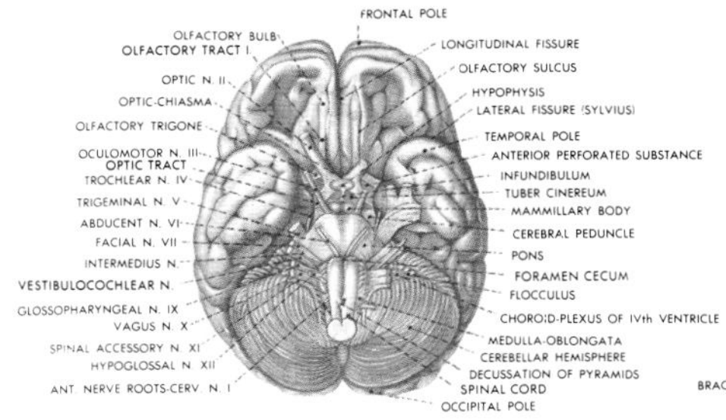

BASE OF THE BRAIN

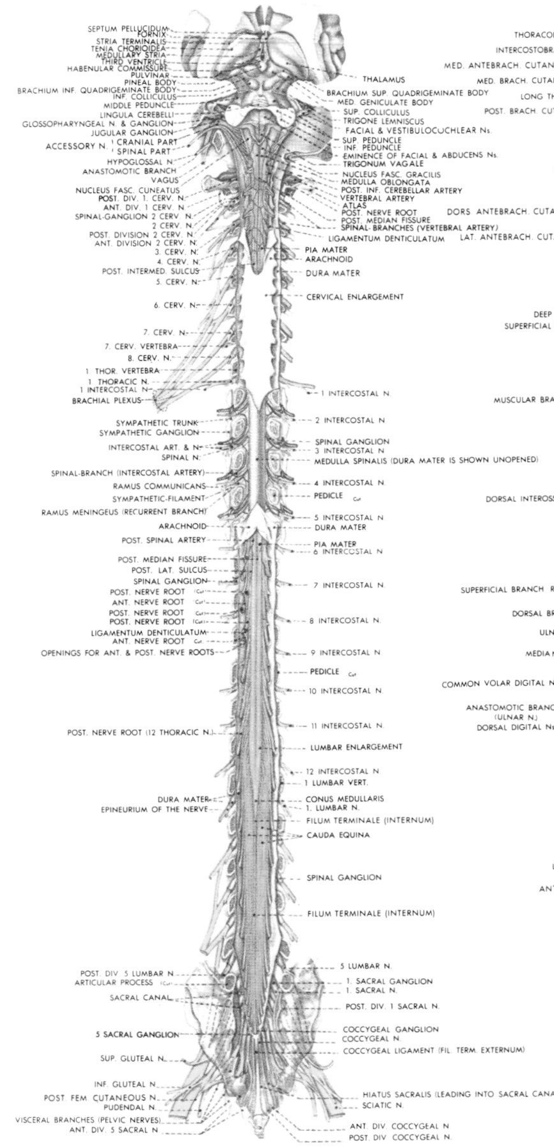

MIDBRAIN-MEDULLA OBLONGATA AND SPINAL CORD

(Posterior View)

Brain, Medulla Oblongata Upper Part of Cervical Region and Lower Half of Medulla Spinalis the Arachnoid Is Removed.

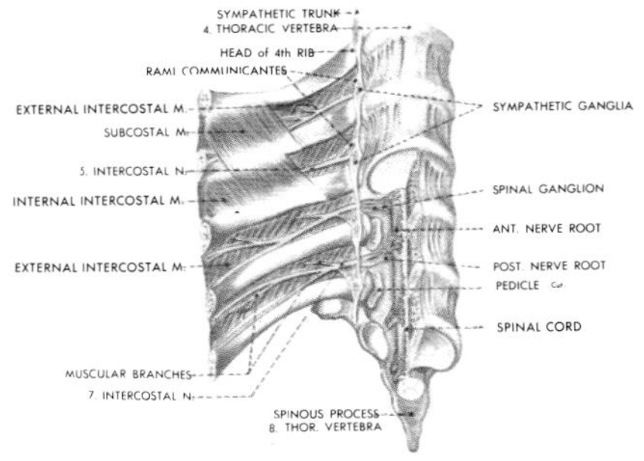

INTERCOSTAL NERVES

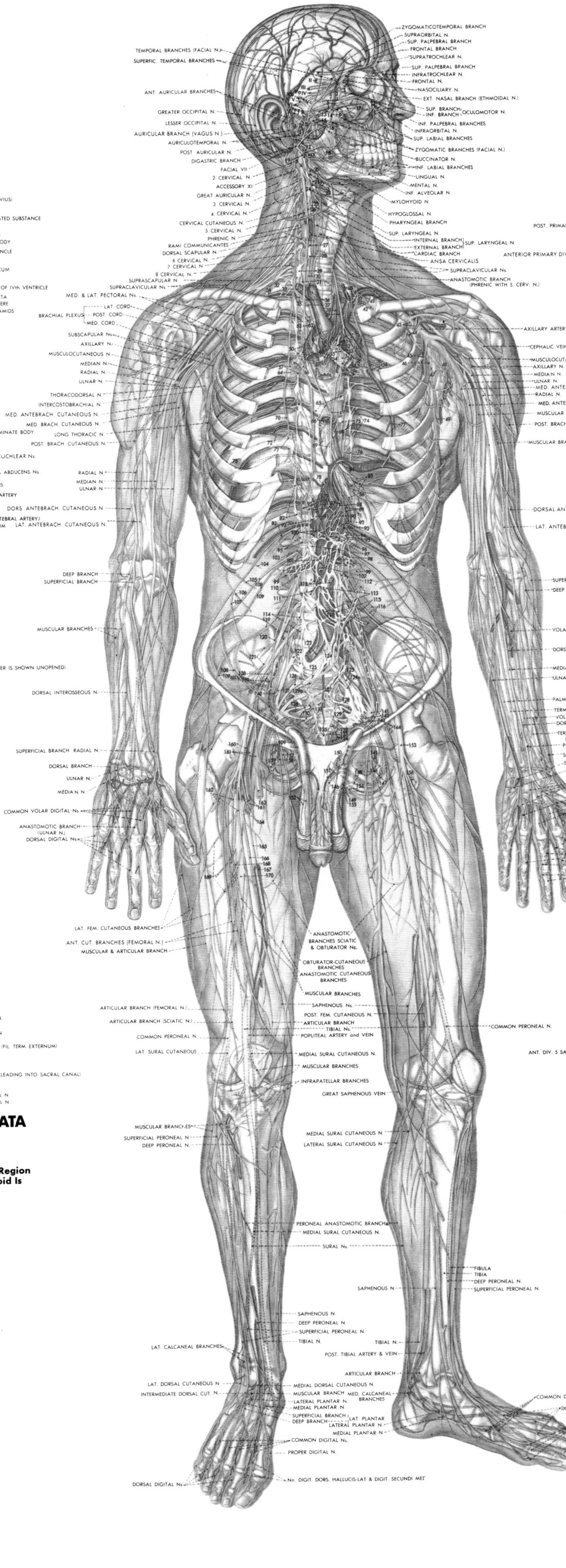

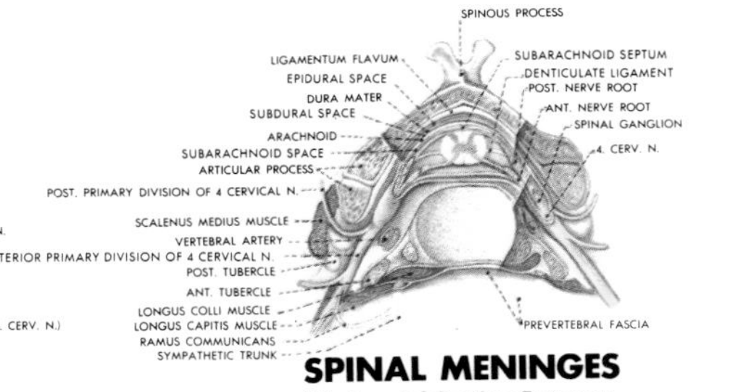

SPINAL MENINGES

(Horizontal Section Between the 4th and 5th Cervical Vertebrae)

HEAD REGION

I OLFACTORY TRACT
II OPTIC N.
III OCULOMOTOR N.
IV TROCHLEAR N.
V TRIGEMINAL N. (OPHTHALMIC N., MAXILLARY N., MANDIBULAR N.)
Va PORTIO MAJOR (SENSORY) } TRIGEMINAL N.
Vb PORTIO MINOR (MOTOR) }
Vc SEMILUNAR GANGLION
VI ABDUCENT N.
VII FACIAL N.
VIII VESTIBULOCOCHLEAR N.
IX GLOSSOPHARYNGEAL N.
X VAGUS N.
XI ACCESSORY N.
XII HYPOGLOSSAL N.
1 CILIARY GANGLION and SHORT CILIARY N.s.
2 LACRIMAL N.
3 ZYGOMATIC N.
4 SPHENOPALATINE GANGLION
5 NERVE of the PTERYGOID CANAL
6 PALATINE N.s.
7 GREATER PETROSAL N.
7a LESSER PETROSAL N.
8 DEEP PETROSAL N. (CAROTID PLEXUS)
9 TYMPANIC N. (NERVE OF JACOBSON) (GLOSSOPHARYNGEAL)
10 INT. CAROTID PLEXUS
11 CHORDA TYMPANI N. (FACIAL)
12 INTERNAL CAROTID N.
13 DEEP TEMPORAL N. (MANDIBULAR)
14 MASSETERIC N.
15 OTIC GANGLION
16 MED. & LAT. PTERYGOID Ns.
17 SUP. ALVEOLAR BRANCHES
18 SUBMANDIBULAR GANGLION

NECK

19 JUGULAR BRANCH
20 PHARYNGEAL BRANCH (GLOSSOPHARYNGEAL)
21 PHARYNGEAL BRANCH (SUP. CERV. GANGLION)
22 EXT. CAROTID-PLEXUS BRANCH
23 PHARYNGEAL PLEXUS
24 SUP. CERV. GANGLION
25 SUP. CARDIAC N.
26 SUP. CARDIAC BRANCH (VAGUS)
27 SYMPATHETIC TRUNK
28 MIDDLE CERV. GANGLION
29 INF. CERV. GANGLION
30 MIDDLE CARDIAC N.
31 INF. CARDIAC N.
32 1 THORACIC GANGLION
33 ANSA-SUBCLAVIA
34 1 INTERCOSTAL N.
35 N. CARDIACUS IMUS
36 RIGHT RECURRENT N.
37 INF. CARDIAC BRANCH (VAGUS)
38 SUP. CARDIAC N.
39 SUP. CARDIAC BRANCH (VAGUS)
40 VERTEBRAL PLEXUS
41 SUBCLAVIAN-PLEXUS
42 SUBCLAVIAN N.

THORAX

43 ANT. THORACIC Ns.
44 INTERCOSTOBRACHIAL N.
45 LAT. THORACIC ART. and VEIN
46 LONG THORACIC N.
47 SUBSCAPULAR Ns.
48 SUBSCAPULAR ART. and VEIN
49 THORACODORSAL ART. and Vein
50 ANT. BRONCHIAL BRANCH
51 LEFT RECURRENT N.
52 LIG. ARTERIOSUM
53 POST. CORONARY PLEXUS
54 ANT. CORONARY PLEXUS
55 CARDIAC GANGLION (WRISBERG)
56 POST. BRONCHIAL BRANCHES
57 POST. PULMONARY PLEXUS
58 ANT. PULMONARY PLEXUS
59 THORACIC AORTIC PLEXUS
60 SUPERF. CARDIAC PLEXUS
61 DEEP CARDIAC PLEXUS
62 TRACHEAL and ESOPHAGEAL BRANCHES
63 PULMONARY BRANCHES
64 PERICARDIAL BRANCH
65 THORACIC AORTIC BRANCH and PLEXUS
66 RIGHT VAGUS N.
67 LEFT VAGUS N.
68 ESOPHAGEAL PLEXUS
69 ESOPHAGEAL BRANCHES
70 RIGHT PHRENIC N.
71 PHRENICO-ABDOMINAL BRANCH
72 BRANCHES OF PHRENIC N.
73 DIAPHRAGM
74 RAMI COMMUNICANTES
75 ROOTS of GREATER SPLANCHNIC N.
76 SYMPATHETIC TRUNK
77 LEFT PHRENIC N. and VESSELS
78 THORACIC AORTA
79 GREATER SPLANCHNIC N.
80 LESSER SPLANCHNIC N.
81 SPLANCHNIC GANGLION
82 THORACIC GANGLIA
83 12 RIB & N.

ABDOMEN

84 ESOPHAGEAL HIATUS
85 ANT. GASTRIC PLEXUS
86 POST. GASTRIC PLEXUS
87 RIGHT VAGUS N.
88 LEFT GASTRIC ARTERY
89 SUP. GASTRIC PLEXUS
90 PHRENIC PLEXUS
91 PHRENIC GANGLION
92 ANT. GASTRIC BRANCH
93 SUPRARENAL BRANCHES
94 LEFT GREATER and LESSER SPLANCHNIC Ns.
95 SUPRARENAL PLEXUS
96 LIENAL PLEXUS
97 CELIAC GANGLIA
98 AORTICORENAL GANGLION
99 RENAL PLEXUS
100 HEPATIC PLEXUS
101 SUP. MESENTERIC GANGLION
102 SUP. MESENTERIC PLEXUS
103 1 LUMBAR N.
104, 105, 106, 107 — 9,10,11,12 INTERCOSTAL Ns.
108 ILIOHYPOGASTRIC N.
109 ILIOINGUINAL N.
110 2 LUMBAR N.
111 TESTICULAR PLEXUS
112 SYMPATHETIC TRUNK
113 ABDOMINAL AORTIC PLEXUS
114 RAMI COMMUNICANTES
115 INF. MESENTERIC PLEXUS
116 SUP. CLUNIAL Ns. (DORSAL LUMBAR BRANCHES 1,2,3 LUMB. N.
117 INF. MESENTERIC ARTERY
118 SPLANCHNIC LUMBAR Ns.
119 LAT. FEMORAL CUTANEOUS N.

PELVIS & PERINEUM

120 GENITOFEMORAL N.
121 MUSCULAR BRANCHES
122 5 LUMBAR N. GANGLION
123 ILIAC PLEXUS
124 SUP. HYPOGASTRIC PLEXUS
125 INF. HYPOGASTRIC PLEXUS
126 1 SACRAL N. GANGLION
127 SUP. RECTAL PLEXUS
128 MIDDLE RECTAL PLEXUS
129 PELVIC NERVES (Nn. ERIGENTES)
129a ASCENDING FIBERS (Nn. ERIGENTES)
130 3 and 4 SACRAL Ns.
131 SYMPATHETIC TRUNK
132 5 SACRAL N.
133 COCCYGEAL N.
134 ANOCOCCYGEAL Ns.
135 GANGLION IMPAR
136 MUSCULAR BRANCHES
137 SUP. GLUTEAL N.
138 LUMBOINGUINAL N.
139 TESTICULAR N.
140 OBTURATOR N.
141 SCIATIC N.
142 POST. FEM. CUTANEOUS N.
143 PUDENDAL N.
144 INF. GLUTEAL N.
145 INF. RECTAL Ns.
146 DORSAL PENIS N.
147 PERINEAL N.
148 PERINEAL BRANCHES
149 POST. SCROTAL Ns.
150 LESSER CAVERNOUS Ns. } PARA-
151 GREATER CAVERNOUS Ns. } SYMPATHETIC
152 ANT. SCROTAL Ns.

LOWER LIMB

(Left)

153 FEMORAL N.
154 PERINEAL BRANCHES (POST. FEM. CUT.)
155 SCROTAL BRANCH (POST. FEM. CUT.)
156 INF. CLUNIAL Ns.

(Right)

157 ANT. BRANCH } OBTURATOR N.
158 POST. BRANCH }
159 ARTICULAR BRANCH
160 MUSCULAR BRANCH
161 FEMORAL ART. and VEIN
162 MUSCULAR BRANCHES
163 GREAT SAPHENOUS VEIN
164 DEEP FEMORAL ART. and VEIN
165 CUTANEOUS BRANCH
166 SAPHENOUS N.
167 FEMORAL ARTERY
168 ARTICULAR BRANCH
169 MUSCULAR BRANCHES
170 SCIATIC N.

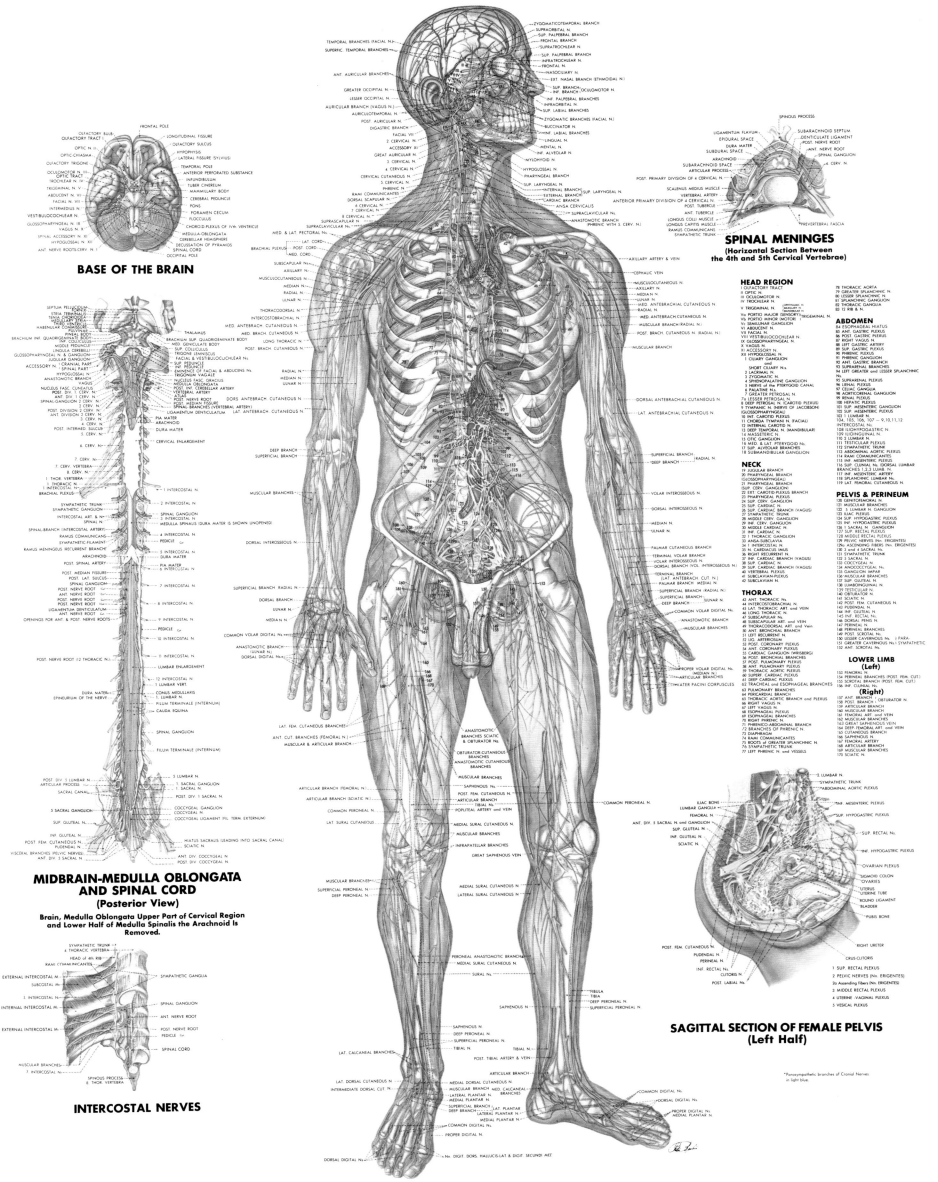

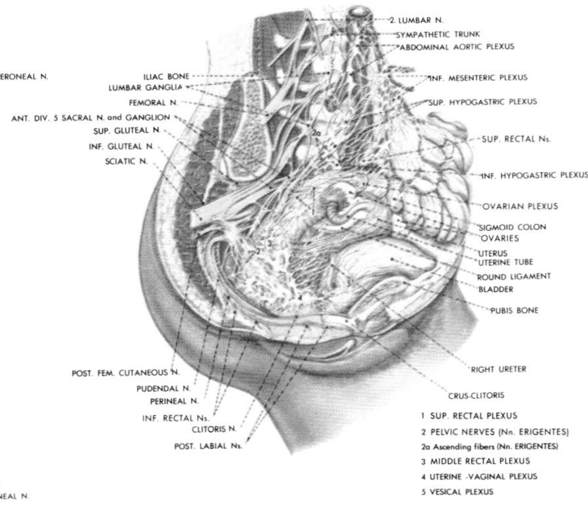

1 SUP. RECTAL PLEXUS
2 PELVIC NERVES (Nn. ERIGENTES)
2a Ascending fibers (Nn. ERIGENTES)
3 MIDDLE RECTAL PLEXUS
4 UTERINE - VAGINAL PLEXUS
5 VESICAL PLEXUS

SAGITTAL SECTION OF FEMALE PELVIS

(Left Half)

*Parasympathetic branches of Cranial Nerves in light blue.

THE LYMPHATIC SYSTEM

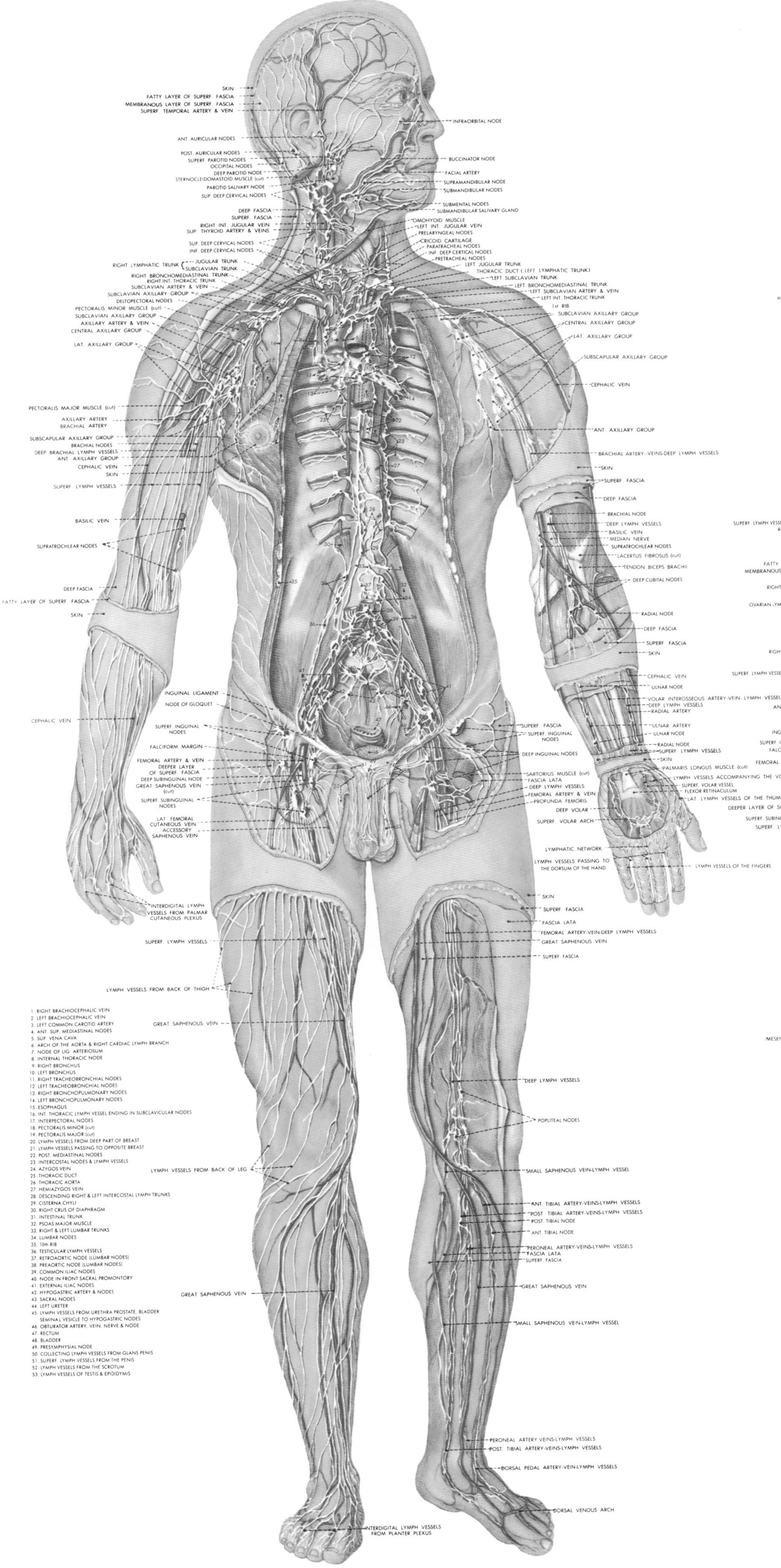

1. RIGHT BRACHIOCEPHALIC VEIN
2. LEFT BRACHIOCEPHALIC VEIN
3. LEFT COMMON CAROTID ARTERY
4. ANT. SUP. MEDIASTINAL NODES
5. SUP. VENA CAVA
6. ARCH OF THE AORTA & RIGHT CARDIAC LYMPH BRANCH
7. NODE OF LIG. ARTERIOSUM
8. INTERNAL THORACIC NODE
9. RIGHT BRONCHUS
10. LEFT BRONCHUS
11. RIGHT TRACHEOBRONCHIAL NODES
12. LEFT TRACHEOBRONCHIAL NODES
13. RIGHT BRONCHOPULMONARY NODES
14. LEFT BRONCHOPULMONARY NODES
15. ESOPHAGUS
16. INT. THORACIC LYMPH VESSEL ENDING IN SUBCLAVICULAR NODES
17. INTERPECTORAL NODES
18. PECTORALIS MINOR (cut)
19. PECTORALIS MAJOR (cut)
20. LYMPH VESSELS FROM DEEP PART OF BREAST
21. LYMPH VESSELS PASSING TO OPPOSITE BREAST
22. POST. MEDIASTINAL NODES
23. INTERCOSTAL NODES & LYMPH VESSELS
24. AZYGOS VEIN
25. THORACIC DUCT
26. THORACIC AORTA
27. HEMIAZYGOS VEIN
28. DESCENDING RIGHT & LEFT INTERCOSTAL LYMPH TRUNKS
29. CISTERNA CHYLI
30. RIGHT CRUS OF DIAPHRAGM
31. INTESTINAL TRUNK
32. PSOAS MAJOR MUSCLE
33. RIGHT & LEFT LUMBAR TRUNKS
34. LUMBAR NODES
35. 10th RIB
36. TESTICULAR LYMPH VESSELS
37. RETROAORTIC NODE (LUMBAR NODES)
38. PREAORTIC NODE (LUMBAR NODES)
39. COMMON ILIAC NODES
40. NODE IN FRONT SACRAL PROMONTORY
41. EXTERNAL ILIAC NODES
42. HYPOGASTRIC ARTERY & NODES
43. SACRAL NODES
44. LEFT URETER
45. LYMPH VESSELS FROM URETHRA PROSTATE, BLADDER SEMINAL VESICLE TO HYPOGASTRIC NODES
46. OBTURATOR ARTERY, VEIN, NERVE & NODE
47. RECTUM
48. BLADDER
49. PRESYMPHYSIAL NODE
50. COLLECTING LYMPH VESSELS FROM GLANS PENIS
51. SUPERF. LYMPH VESSELS FROM THE PENIS
52. LYMPH VESSELS FROM THE SCROTUM
53. LYMPH VESSELS OF TESTIS & EPIDIDYMIS

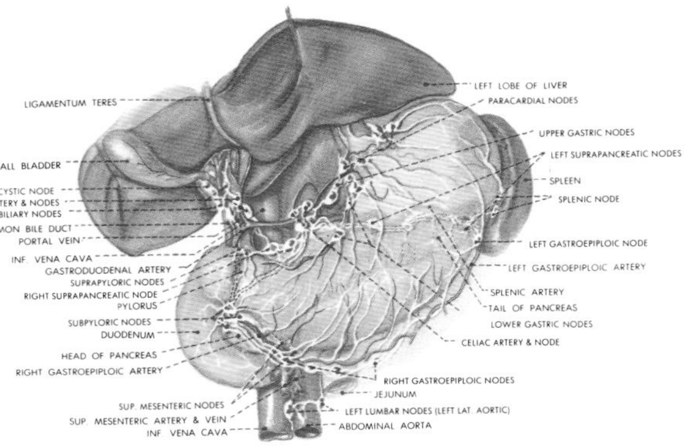

LYMPH-VESSELS and LYMPH NODES OF STOMACH-PANCREAS-SPLEEN and BILIARY TRACT

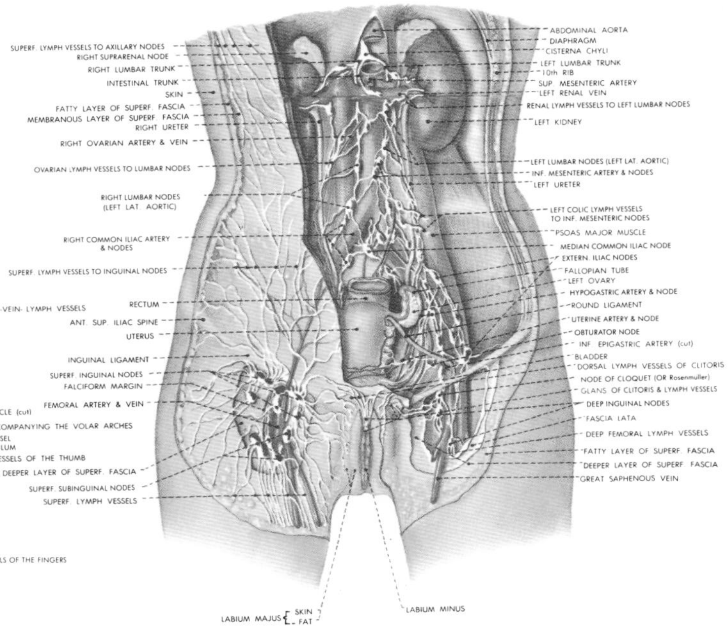

LYMPH-VESSELS AND LYMPH-NODES OF INGUINAL REGION-PELVIS AND POSTERIOR WALL OF ABDOMEN

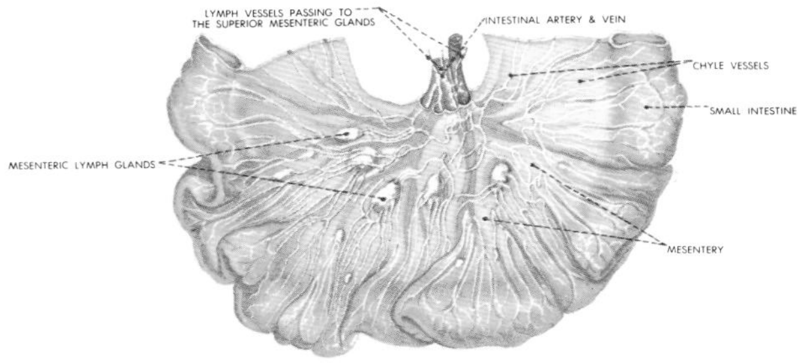

LYMPH VESSELS and LYMPH NODES OF THE SMALL INTESTINE

LYMPH VESSELS PASSING TO THE SUPERIOR MESENTERIC NODES

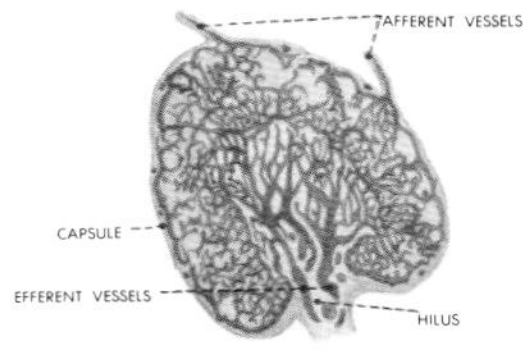

INTERNAL ILIAC LYMPH NODE
(Transverse Section)

THE DIGESTIVE SYSTEM

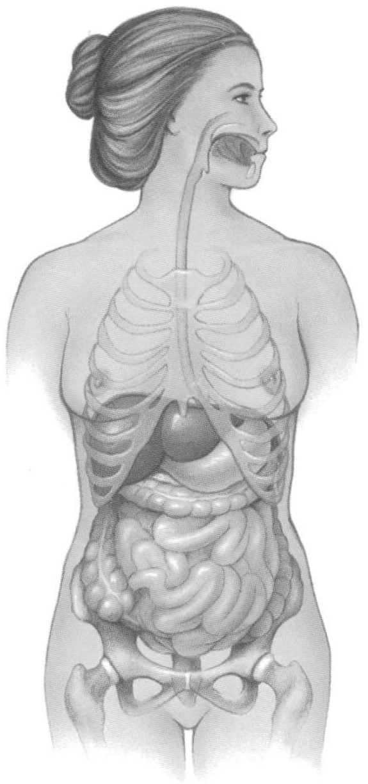

The Oral Cavity, Salivary Glands and Stomach

Digestion begins in the mouth as food is mixed with saliva and broken down into carbohydrates and simpler sugars. After moving to the stomach through the esophagus, food is further broken down by enzymes and hydrochloric acid. A layer of mucus protects the stomach lining from damage by the hydrochloric acid.

Oral Cavity

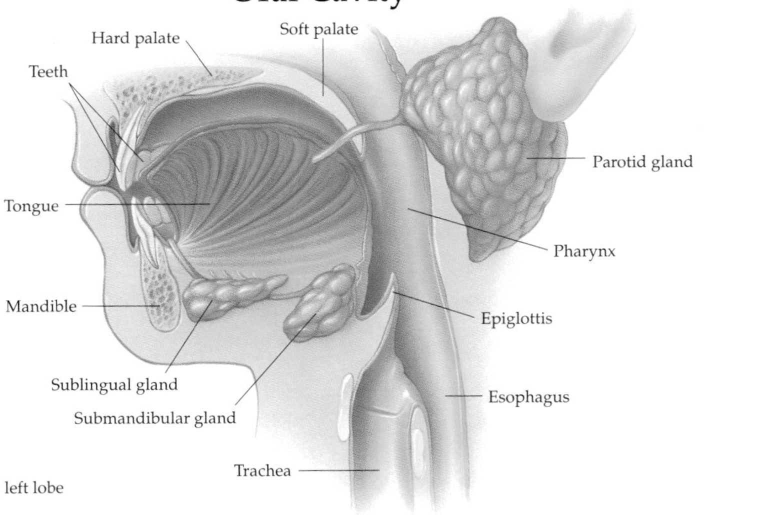

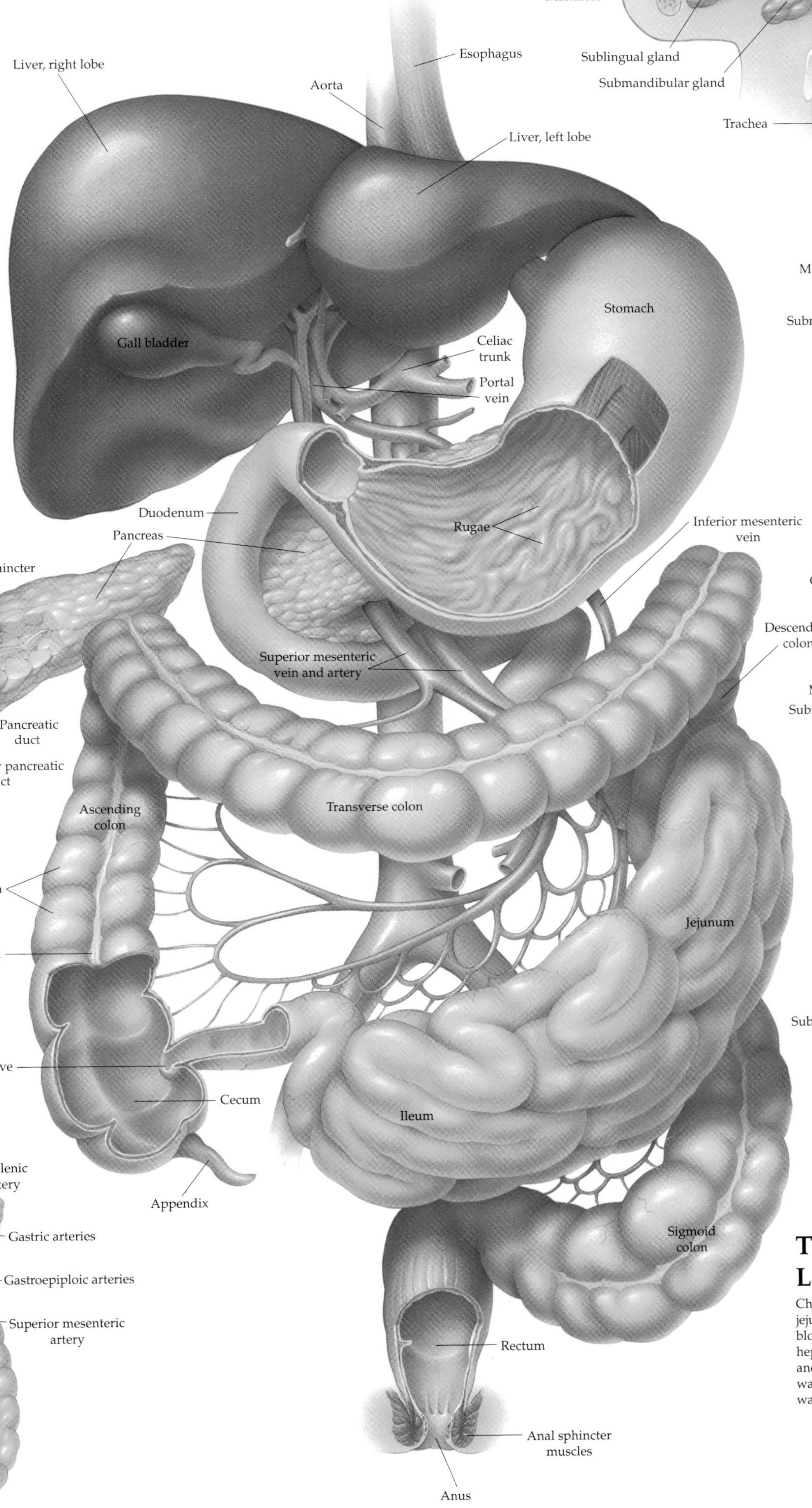

The Liver, Pancreas, and Duodenum

Partially digested food, or chyme, passes from the stomach to the duodenum. Here enzymes, produced in the pancreas, and bile, produced by the liver and stored in the gall bladder, enter the duodenum and further break down fat, protein and carbohydrates.

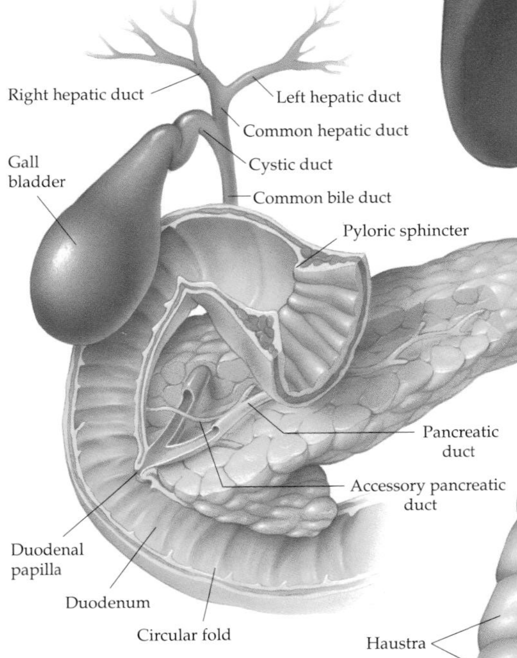

Wall of Stomach

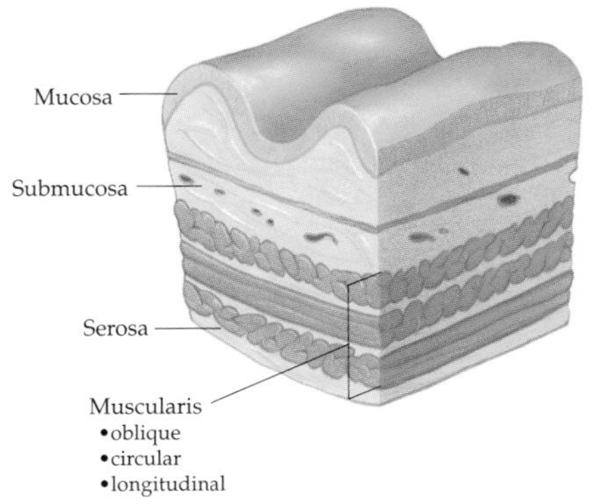

Wall of Jejunum

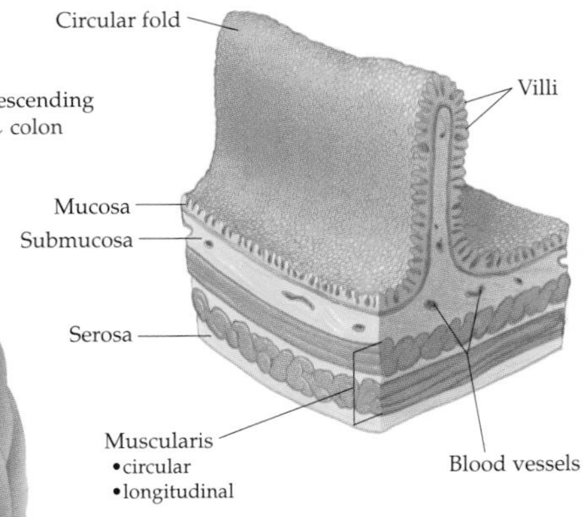

Wall of Colon

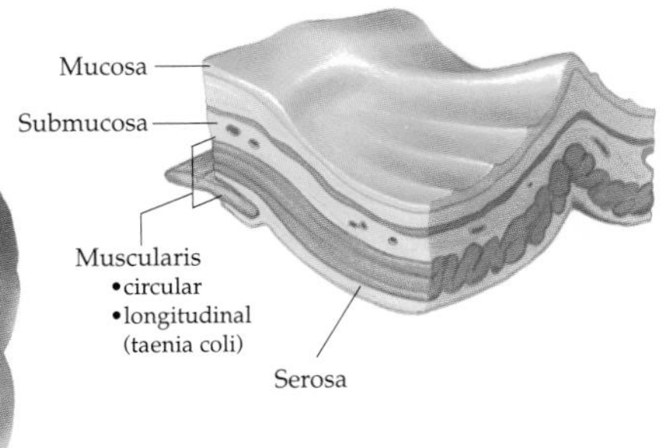

Arterial Supply

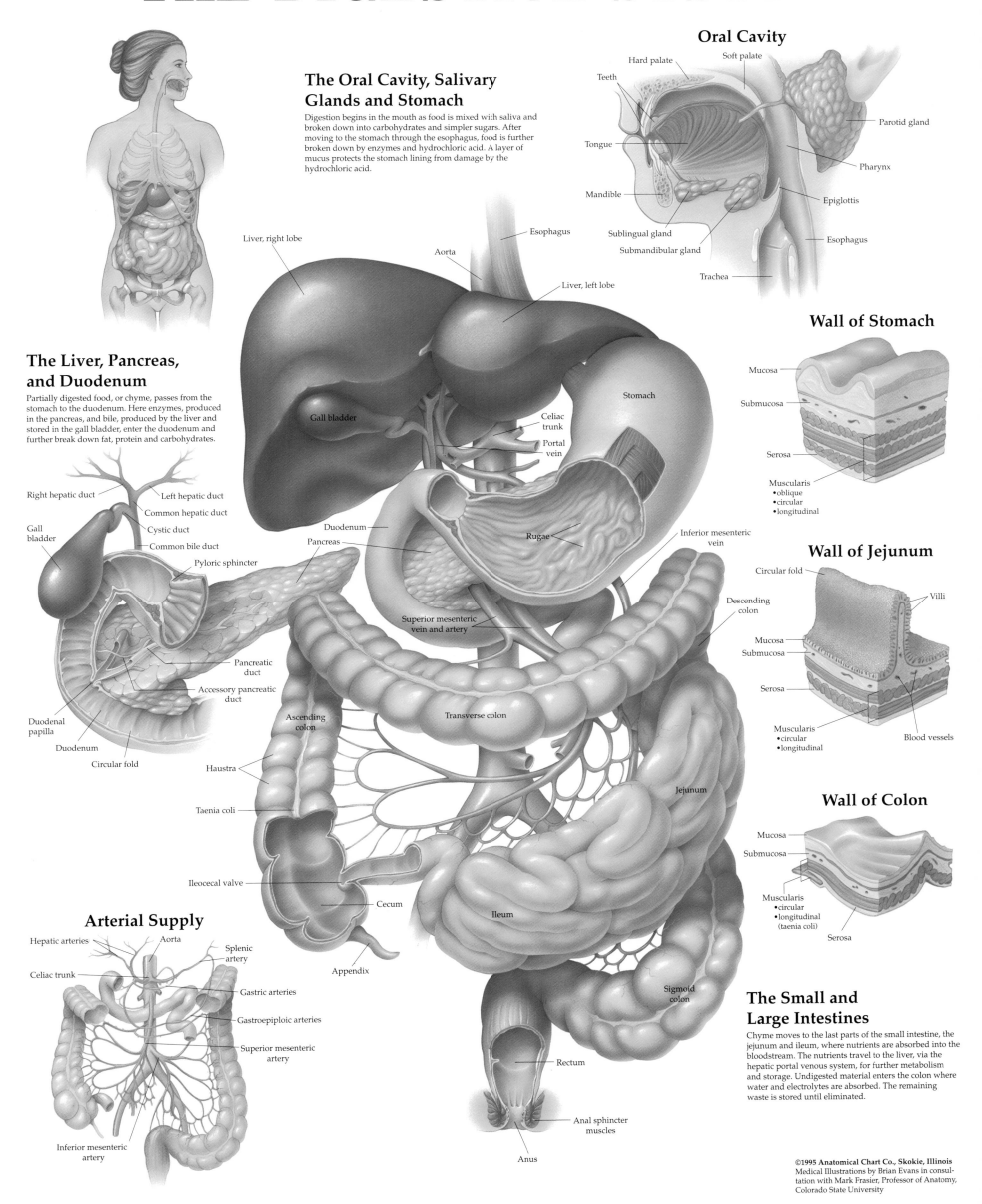

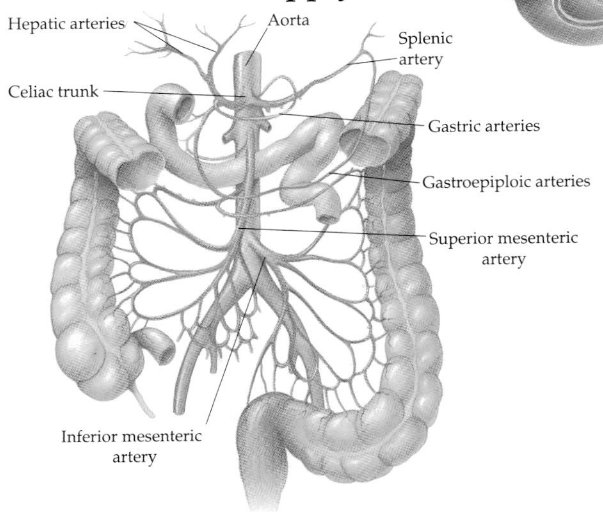

The Small and Large Intestines

Chyme moves to the last parts of the small intestine, the jejunum and ileum, where nutrients are absorbed into the bloodstream. The nutrients travel to the liver, via the hepatic portal venous system, for further metabolism and storage. Undigested material enters the colon where water and electrolytes are absorbed. The remaining waste is stored until eliminated.

Medical Illustrations by Brian Evans in consultation with Mark Frasier, Professor of Anatomy, Colorado State University

THE RESPIRATORY SYSTEM

PARANASAL SINUSES

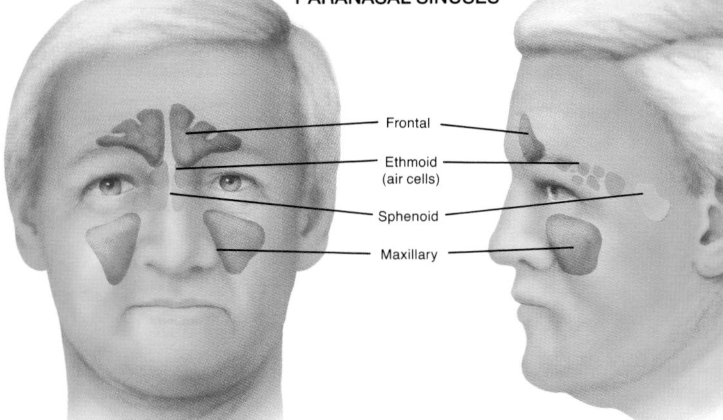

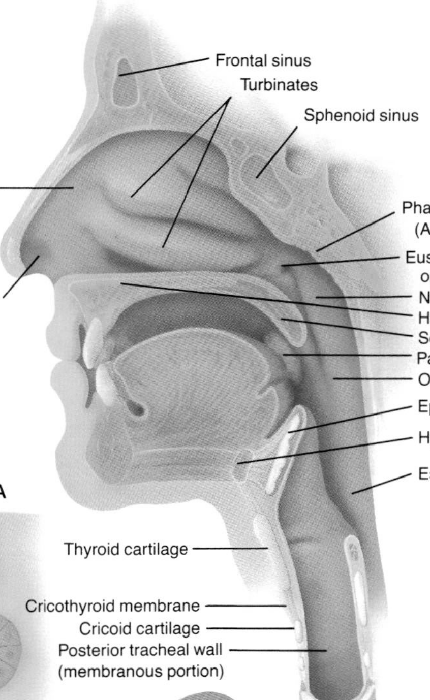

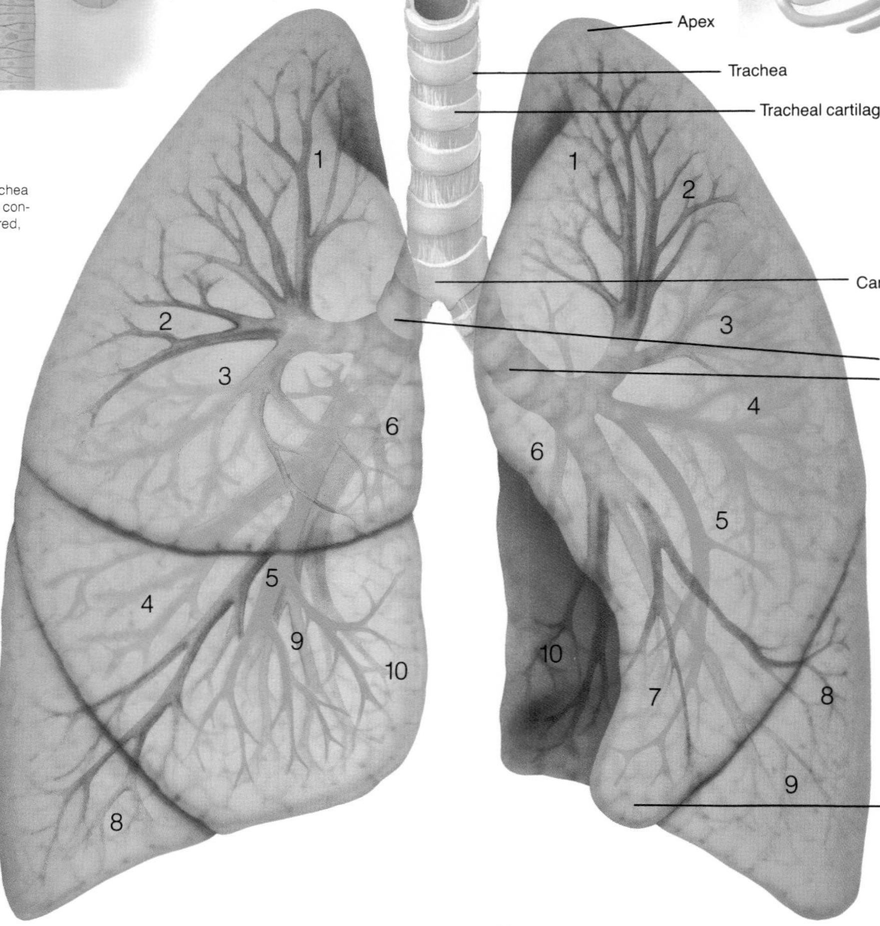

VENTILATION

During breathing, air is inhaled and exhaled, allowing exchange between the atmosphere and the lungs. The diaphragm is the major inspiratory muscle. As the diaphragm contracts, the volume of the thoracic cage increases, causing air flow into the lungs. As the diaphragm relaxes, the elastic tissues of the lung passively recoil, decreasing lung volume.

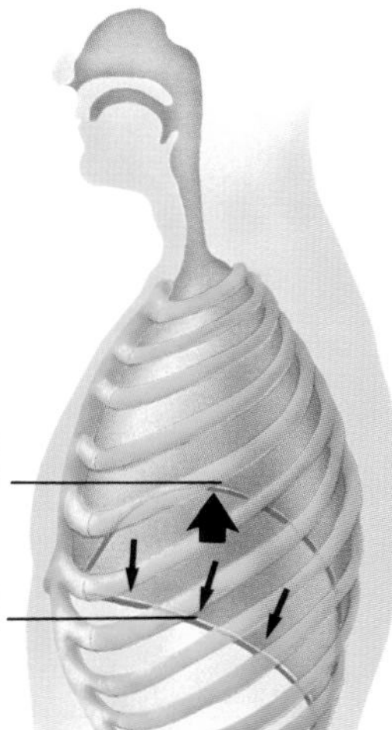

RESPIRATORY MUCOSA

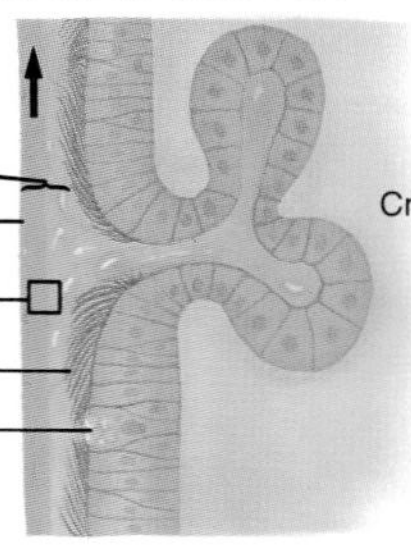

THE LARYNX

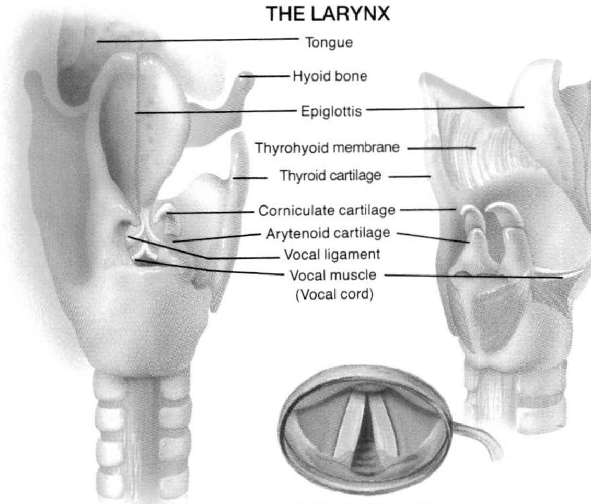

CONDUCTING SYSTEM

The nasal cavity, pharynx, larynx, trachea and bronchial tree form the respiratory conducting pathways. Air is warmed, filtered, moistened and delivered to and from the gas exchange area of the lung.

RESPIRATORY ZONE

The primary lobule (respiratory bronchiole, alveolar duct and alveolar sac) is the functional unit of the respiratory system where oxygen is provided to pulmonary arterial blood and carbon dioxide is removed from the blood. The thin, moist alveolar walls permit gas exchange between the alveolus and the capillary meshwork.

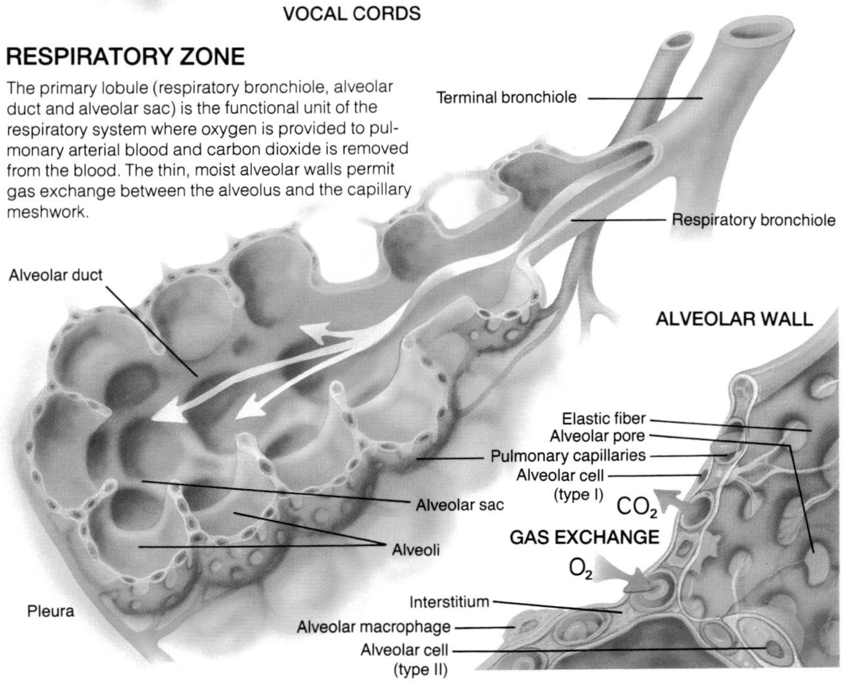

A cigarette smoker's lungs

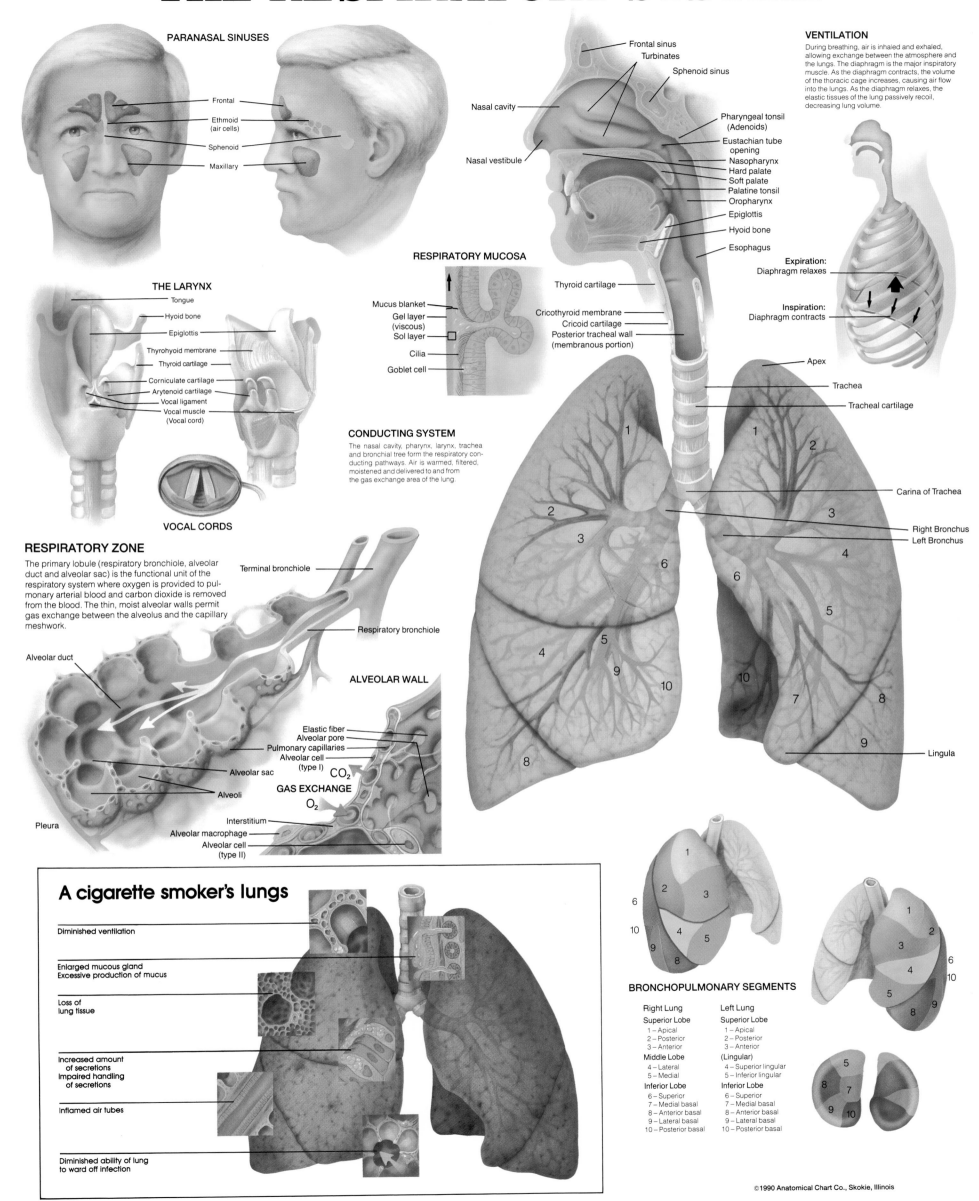

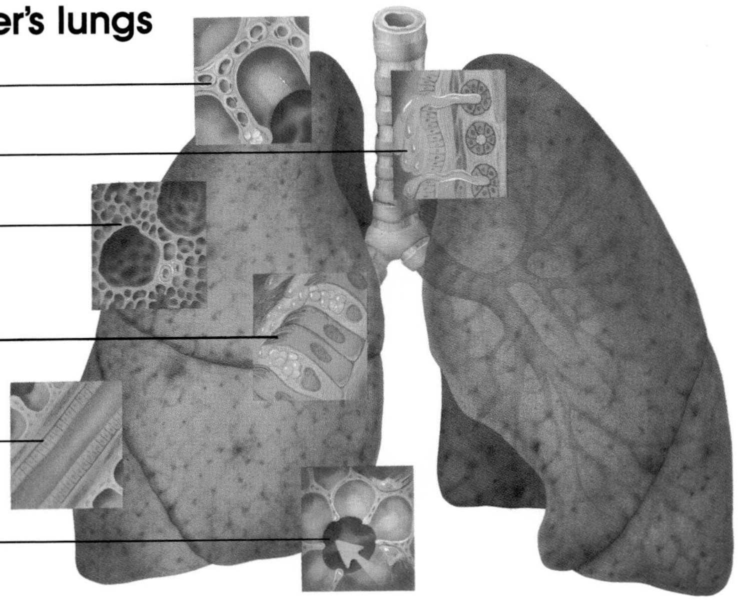

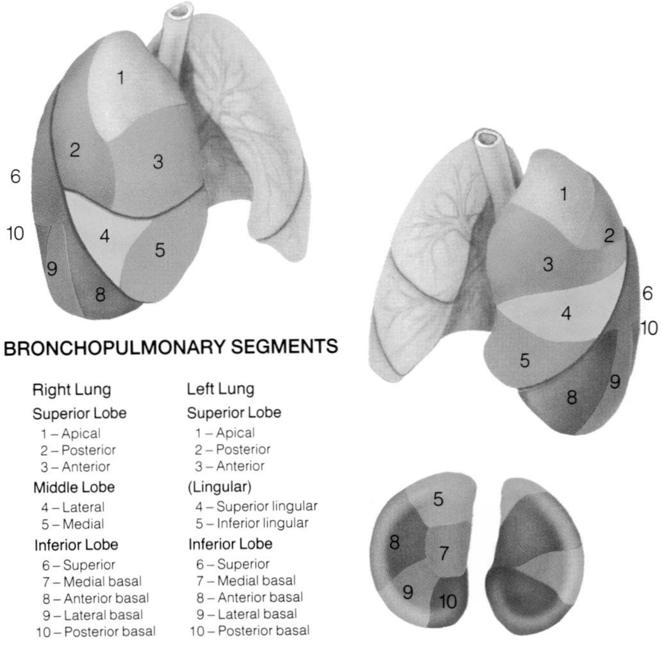

BRONCHOPULMONARY SEGMENTS

Right Lung	Left Lung
Superior Lobe	Superior Lobe
1 – Apical	1 – Apical
2 – Posterior	2 – Posterior
3 – Anterior	3 – Anterior
Middle Lobe	(Lingular)
4 – Lateral	4 – Superior lingular
5 – Medial	5 – Inferior lingular
Inferior Lobe	Inferior Lobe
6 – Superior	6 – Superior
7 – Medial basal	7 – Medial basal
8 – Anterior basal	8 – Anterior basal
9 – Lateral basal	9 – Lateral basal
10 – Posterior basal	10 – Posterior basal

THE URINARY TRACT

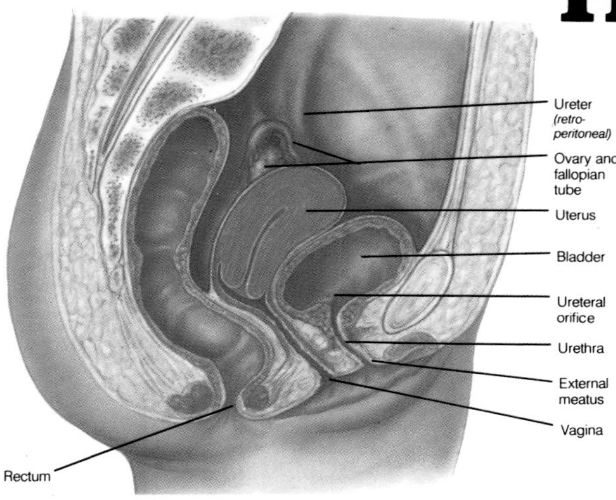

THE FEMALE GENITOURINARY SYSTEM
(x-section)

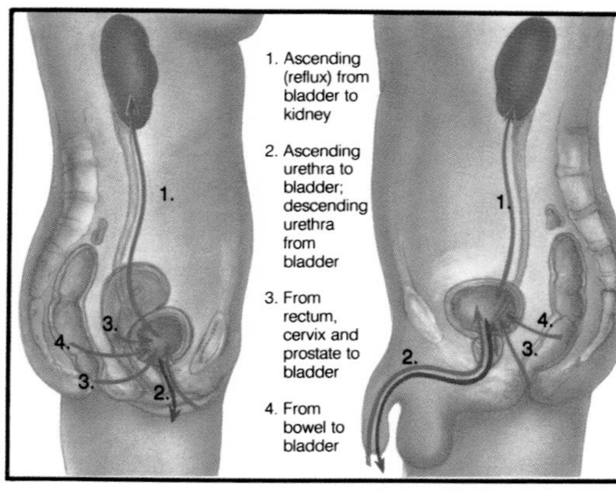

ROUTES OF INFECTION IN THE URINARY TRACT

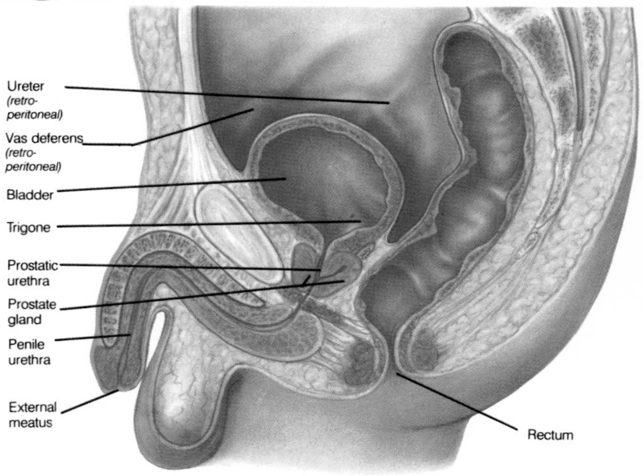

THE MALE GENITOURINARY SYSTEM
(x-section)

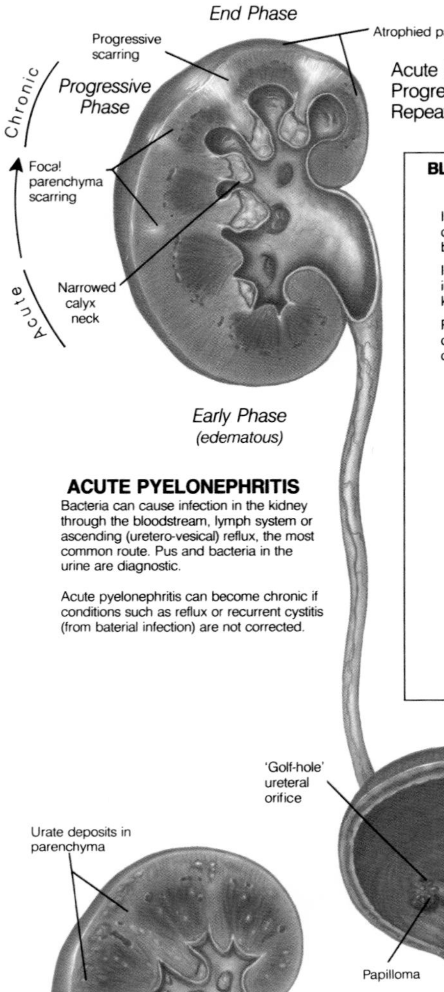

Acute Pyelonephritis and Progressive Scarring From Repeated Infection

BLADDER INFECTION
(Cystitis)

In women, is commonly caused by ascent of bacteria in the urethra.

In men, is caused by infection in the prostate, kidney or residual urine.

Presence of a stone is often complicated by cystitis.

Endoscopic View, Bladder Wall

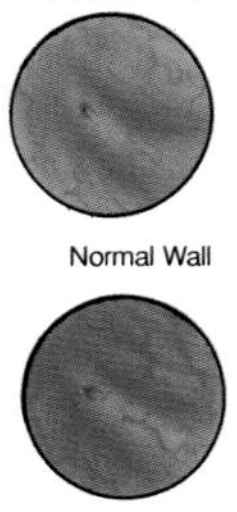

ACUTE PYELONEPHRITIS

Bacteria can cause infection in the kidney through the bloodstream, lymph system or ascending (uretero-vesical) reflux, the most common route. Pus and bacteria in the urine are diagnostic.

Acute pyelonephritis can become chronic if conditions such as reflux or recurrent cystitis (from baterial infection) are not corrected.

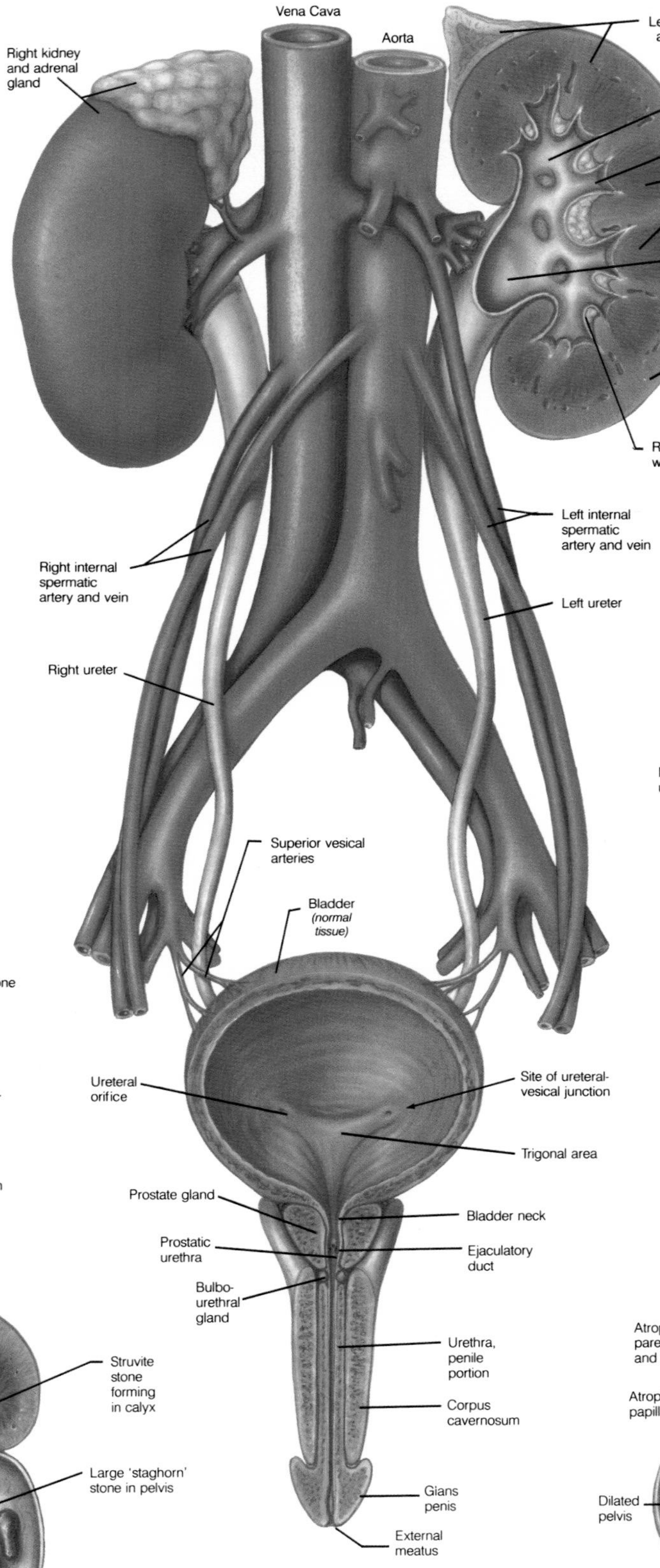

KIDNEYS AND URINARY TRACT

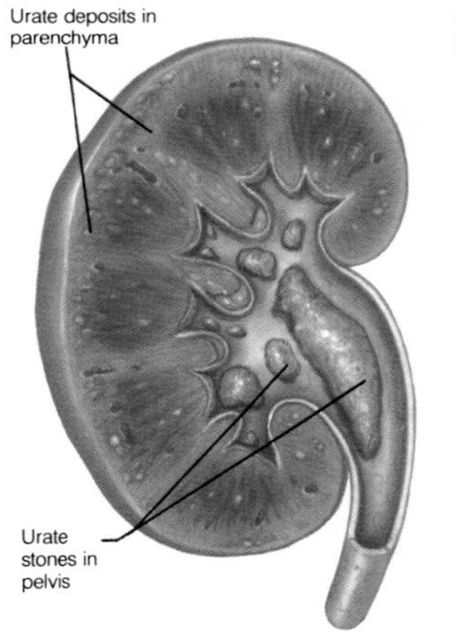

Uric Acid Stones:
Often seen with gout, dehydration, uricosuric drugs, chronic diarrhea, ileostomies and glycogen storage disease.

URINARY STONES

All causes of stone formation are not known, but hyperexcretion of relatively insoluble urinary components such as calcium and phosphate and increased concentration of salts and organic compounds certainly influence stone formation.

The size and position of the stone determine the development of secondary pathologic changes in the urinary tract. Location can be in the kidney, ureter, bladder and urethra (less common).

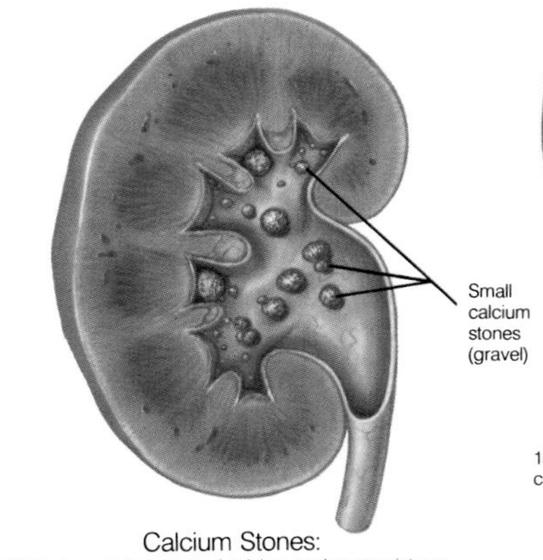

Calcium Stones:
70% of renal stones are of calcium-oxalate or mixtures of calcium-oxalate and calcium-phosphate in the form of hypoxyapatite. Two-thirds of patients with primary hyperparathyroidism have calcium stones.

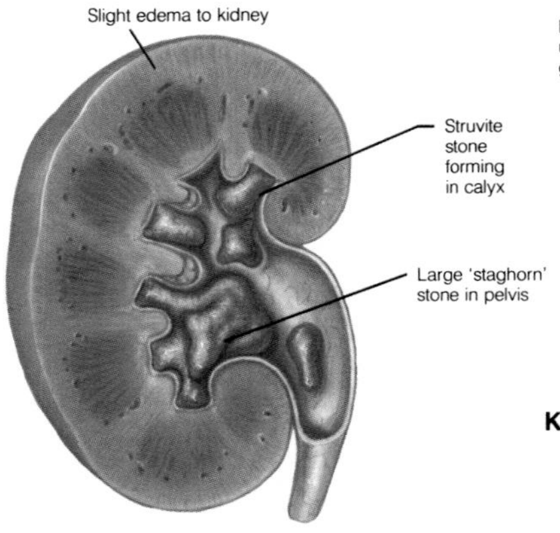

Magnesium-Ammonium Phosphate (Struvite) Stones:
15% of renal stones are of struvite. 'Staghorn' conformations are common.

Vesical Stones
(calcium oxalate)

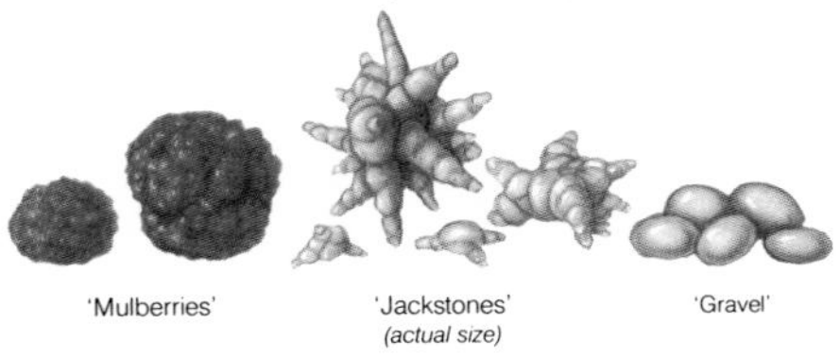

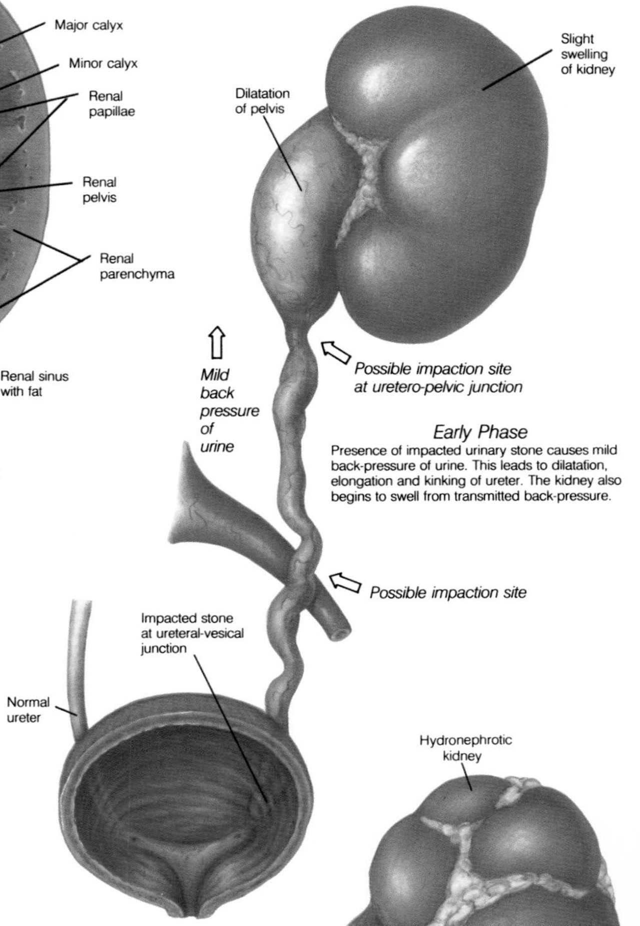

Early Phase
Presence of impacted urinary stone causes mild back-pressure of urine. This leads to dilatation, elongation and kinking of ureter. The kidney also begins to swell from transmitted back-pressure.

URINARY OBSTRUCTION AND STASIS

Obstructions anywhere along the urinary tract, such as an impacted stone, often leads to dilatation and distension of the ureters and renal pelvis.

Continuous increase in intrapelvic pressure can cause ischemia and eventual destruction of kidney tubules, and parenchyma (hydronephrosis).

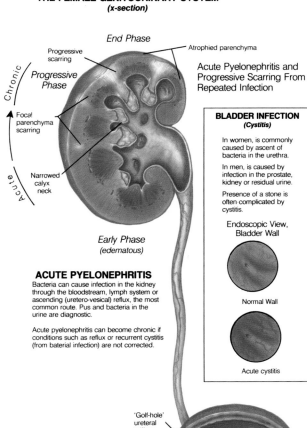

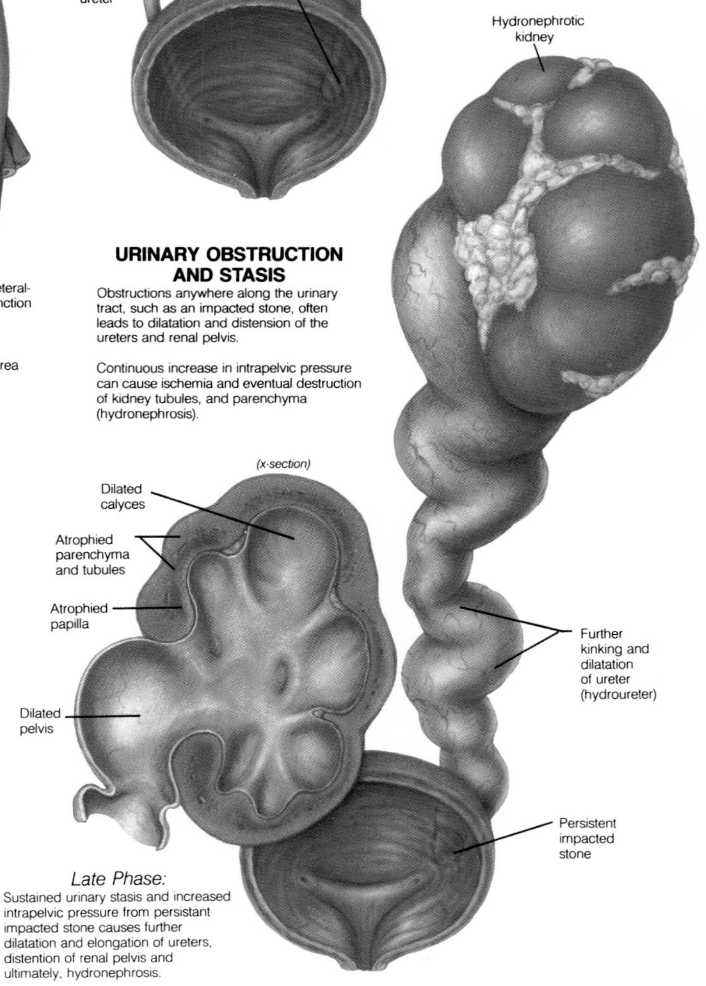

Late Phase:
Sustained urinary stasis and increased intrapelvic pressure from persistant impacted stone causes further dilatation and elongation of ureters, distention of renal pelvis and ultimately, hydronephrosis.

THE FEMALE REPRODUCTIVE SYSTEM

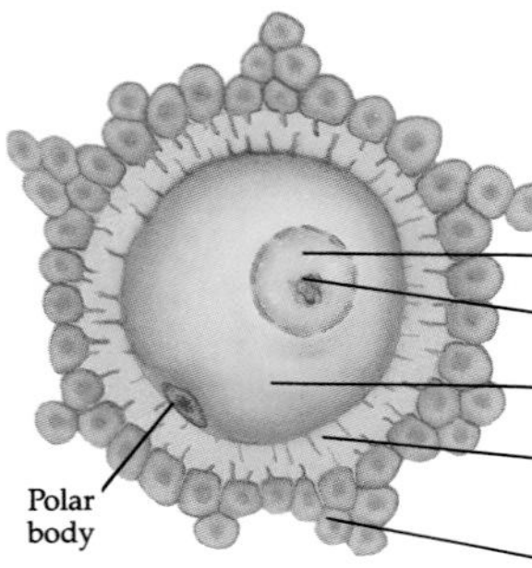

Ovum
(Unfertilized)
100-150um diameter

Ovary, Fallopian Tube, Uterus and Vagina

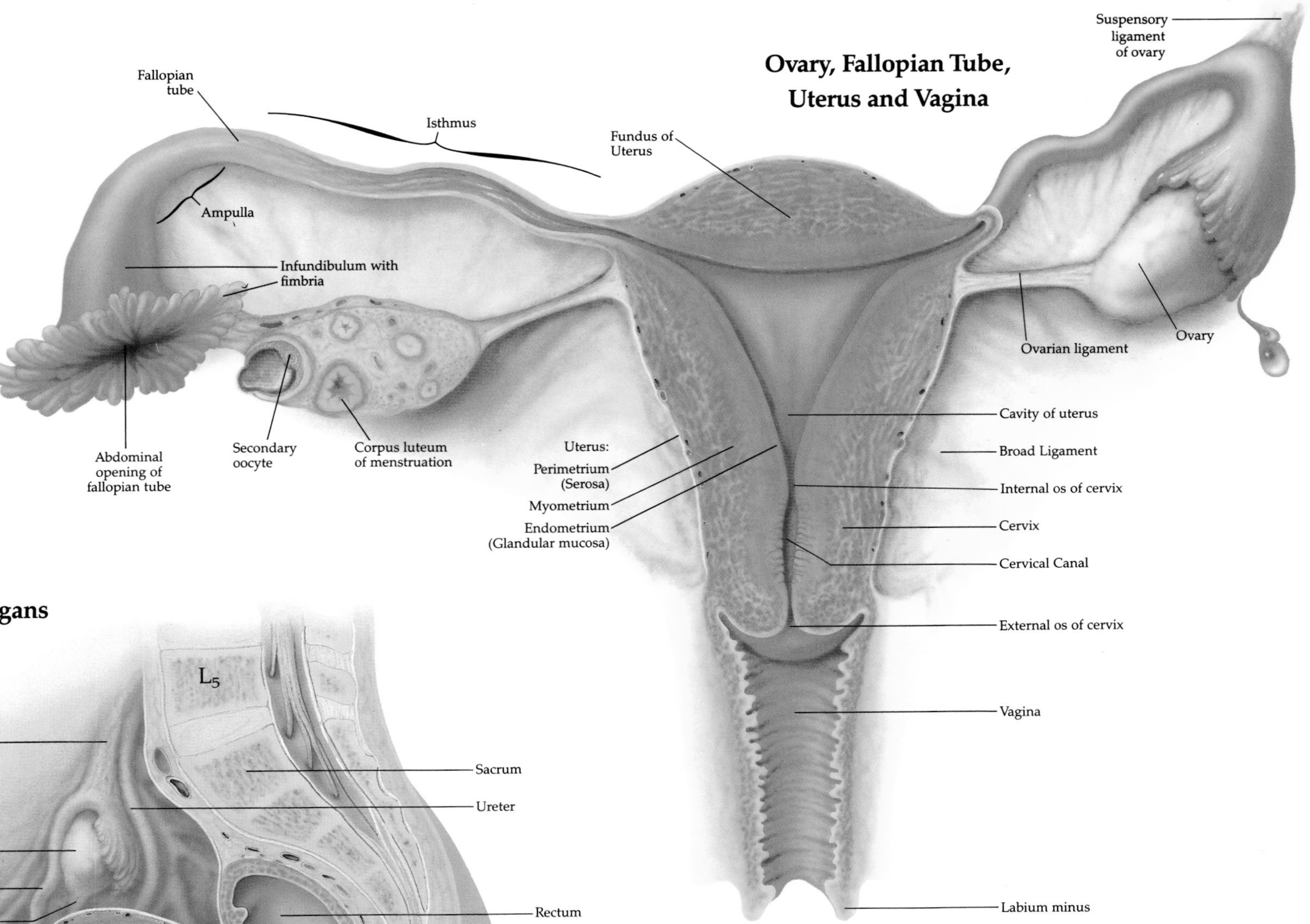

The Female Pelvic Organs
(Median section)

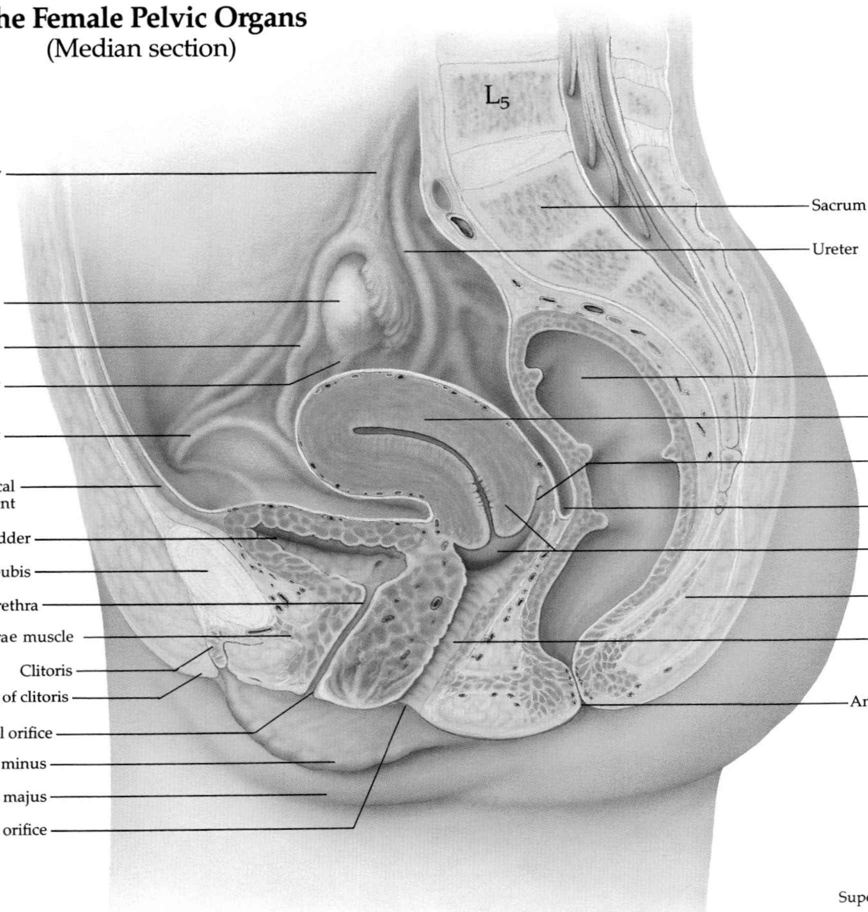

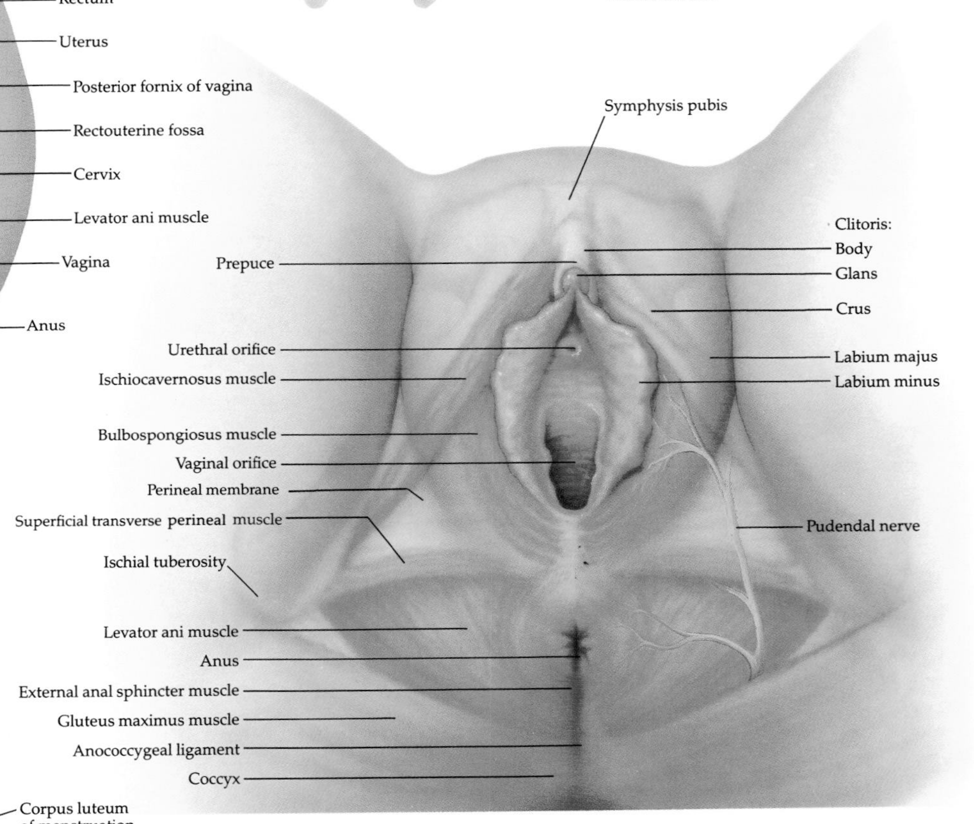

The Female Perineum

The Menstrual Cycle

The menstrual cycle occurs during the reproductive period from puberty through menopause in response to rhythmic variations of hormones. The endometrial lining of the uterus proliferates in preparation for implantation of a fertilized egg and in the absence of pregnancy is shed with some bleeding through the vagina.

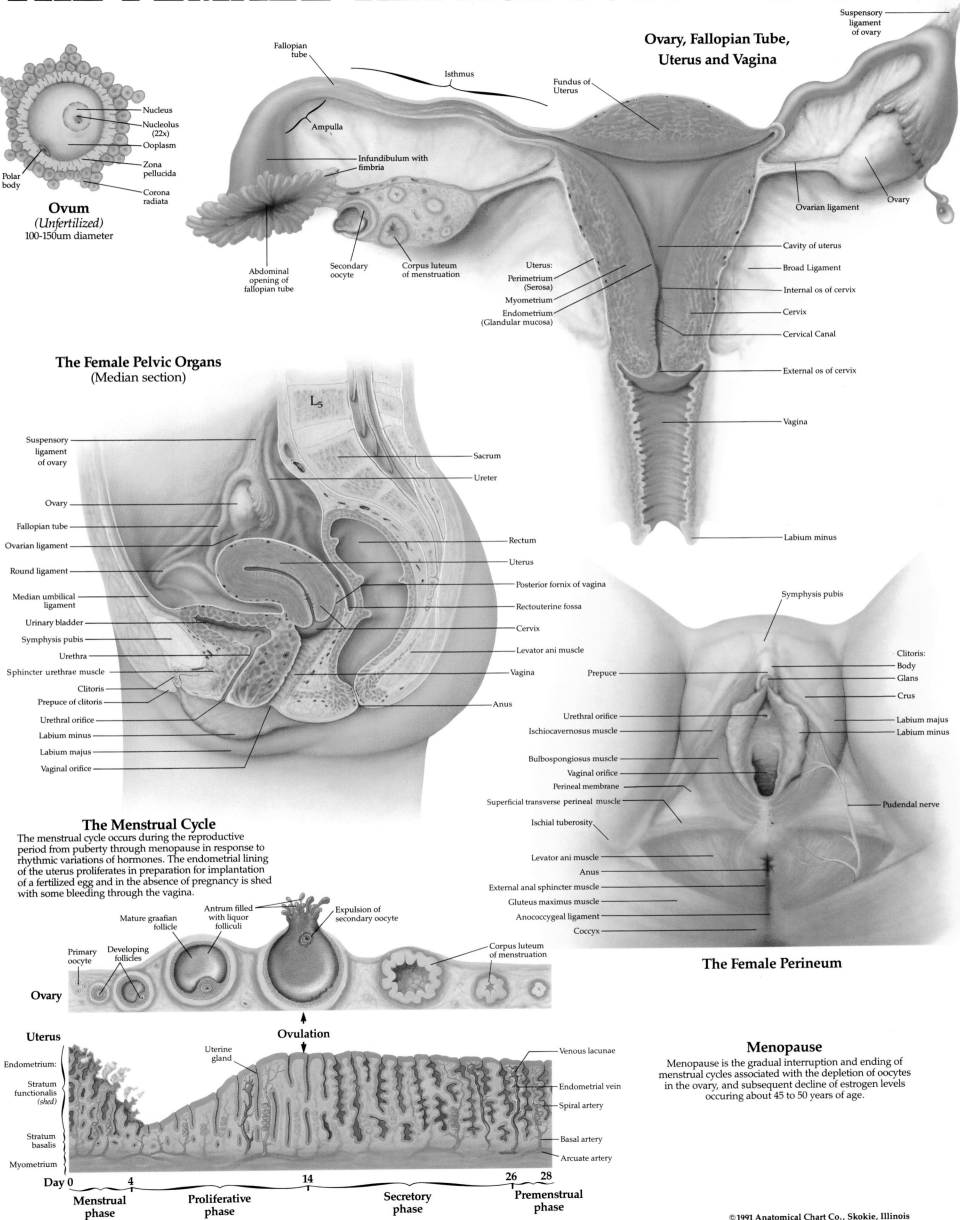

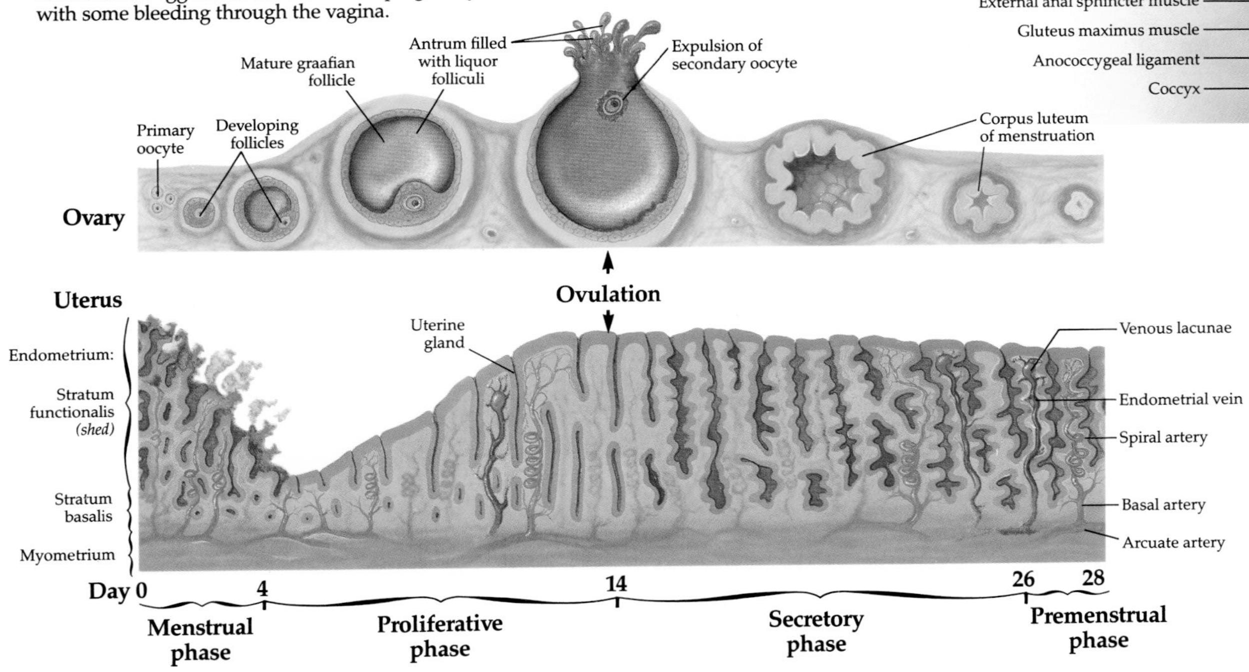

Menopause

Menopause is the gradual interruption and ending of menstrual cycles associated with the depletion of oocytes in the ovary, and subsequent decline of estrogen levels occuring about 45 to 50 years of age.

THE MALE REPRODUCTIVE SYSTEM

Anterior View (Oblique section)

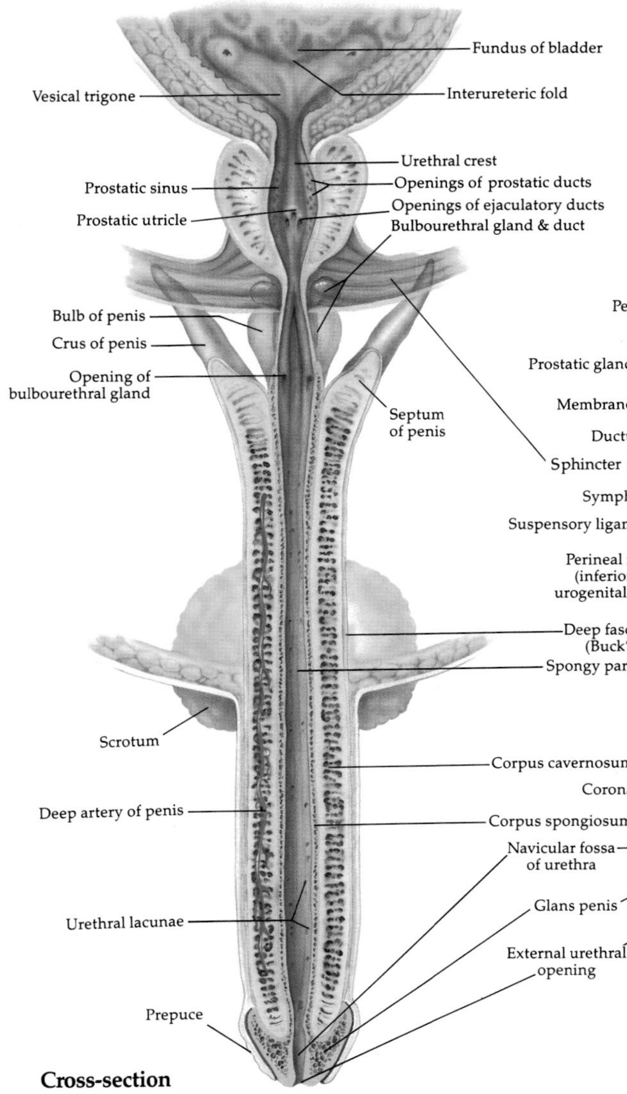

Fundus of bladder
Vesical trigone
Interureteric fold
Urethral crest
Prostatic sinus
Openings of prostatic ducts
Openings of ejaculatory ducts
Prostatic utricle
Bulbourethral gland & duct
Bulb of penis
Crus of penis
Opening of bulbourethral gland
Septum of penis
Scrotum
Deep artery of penis
Urethral lacunae
Prepuce

The Pelvic Organs
(Median section)

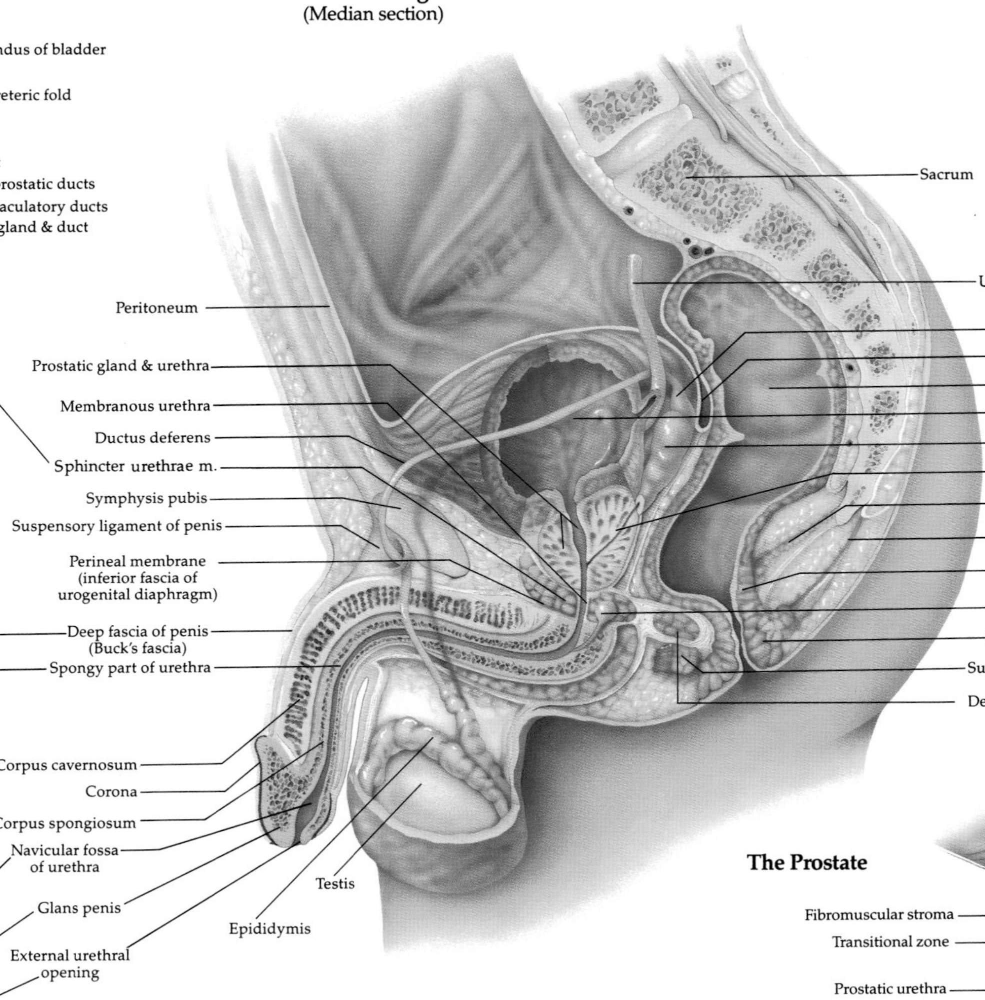

Sacrum
Ureter
Peritoneum
Ampulla of ductus deferens
Rectovesical pouch
Prostatic gland & urethra
Rectum
Membranous urethra
Urinary bladder
Ductus deferens
Seminal vesicle
Sphincter urethrae m.
Ejaculatory duct
Symphysis pubis
Levator ani m.
Suspensory ligament of penis
Anococcygeal ligament
Perineal membrane (inferior fascia of urogenital diaphragm)
Internal anal sphincter m.
Bulbourethral gland & duct
Deep fascia of penis (Buck's fascia)
External anal sphincter m.
Spongy part of urethra
Superficial transverse perineal m.
Deep transverse perineal m.
Corpus cavernosum
Corona
Corpus spongiosum
Navicular fossa of urethra
Testis
Glans penis
Epididymis
External urethral opening

The Prostate

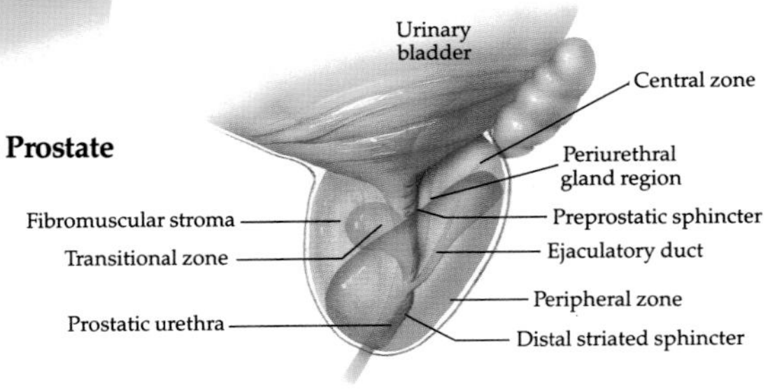

Urinary bladder
Central zone
Periurethral gland region
Fibromuscular stroma
Preprostatic sphincter
Transitional zone
Ejaculatory duct
Peripheral zone
Prostatic urethra
Distal striated sphincter

Cross-section of the Penis

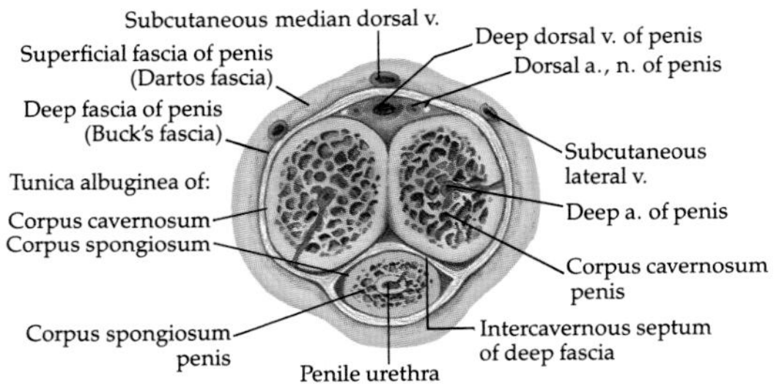

Subcutaneous median dorsal v.
Deep dorsal v. of penis
Superficial fascia of penis (Dartos fascia)
Dorsal a., n. of penis
Deep fascia of penis (Buck's fascia)
Subcutaneous lateral v.
Tunica albuginea of:
Corpus cavernosum
Corpus spongiosum
Deep a. of penis
Corpus cavernosum penis
Corpus spongiosum penis
Intercavernous septum of deep fascia
Penile urethra

Posterior View

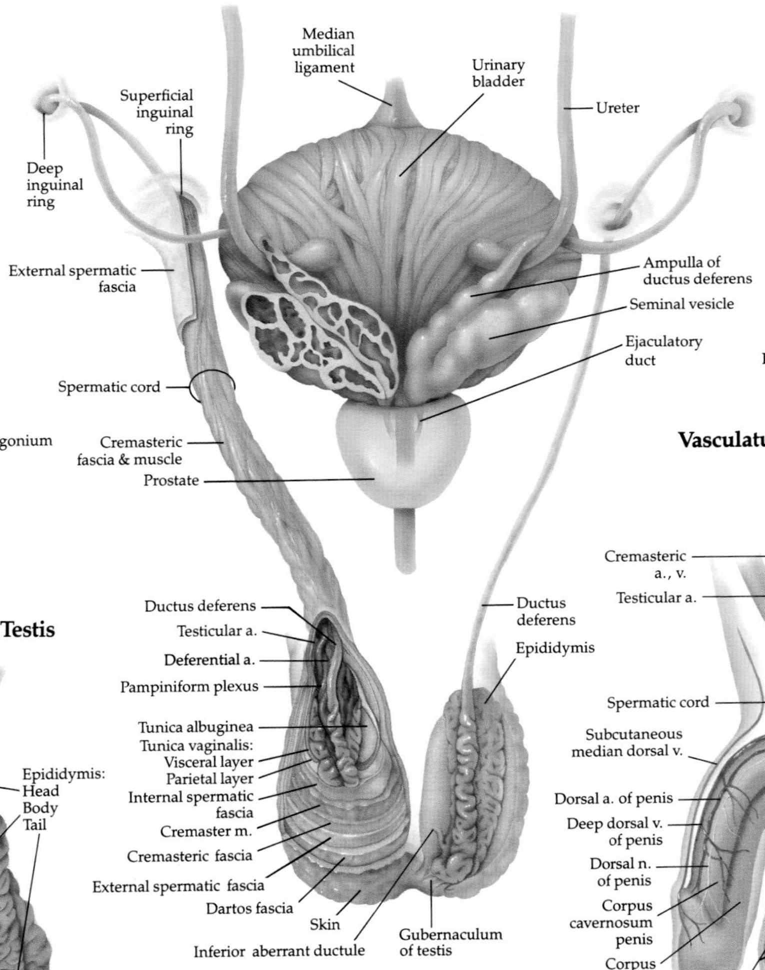

Median umbilical ligament
Urinary bladder
Superficial inguinal ring
Ureter
Deep inguinal ring
Ampulla of ductus deferens
External spermatic fascia
Seminal vesicle
Ejaculatory duct
Spermatic cord
Cremasteric fascia & muscle
Prostate
Ductus deferens
Testicular a.
Deferential a.
Pampiniform plexus
Ductus deferens
Epididymis
Tunica albuginea
Tunica vaginalis:
Visceral layer
Parietal layer
Internal spermatic fascia
Cremaster m.
Cremasteric fascia
External spermatic fascia
Dartos fascia
Skin
Inferior aberrant ductule
Gubernaculum of testis

The Perineum

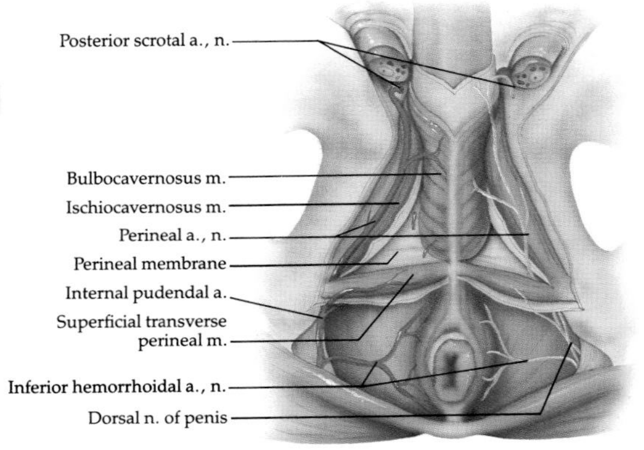

Posterior scrotal a., n.
Bulbocavernosus m.
Ischiocavernosus m.
Perineal a., n.
Perineal membrane
Internal pudendal a.
Superficial transverse perineal m.
Inferior hemorrhoidal a., n.
Dorsal n. of penis

Spermatogenesis

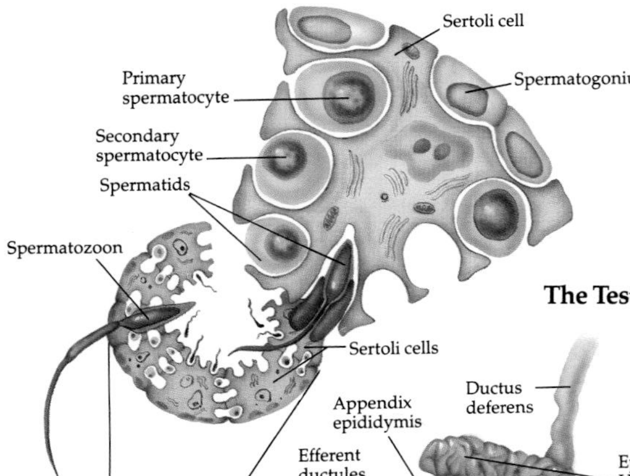

Sertoli cell
Primary spermatocyte
Spermatogonium
Secondary spermatocyte
Spermatids
Spermatozoon
Sertoli cells

The Testis

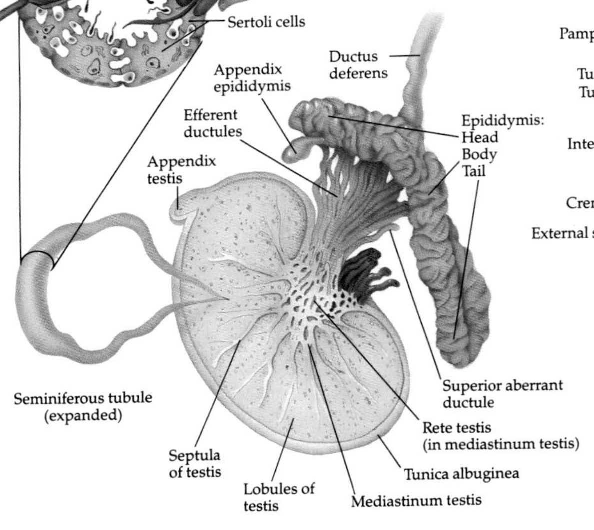

Ductus deferens
Appendix epididymis
Efferent ductules
Epididymis:
Head
Body
Tail
Appendix testis
Seminiferous tubule (expanded)
Superior aberrant ductule
Rete testis (in mediastinum testis)
Tunica albuginea
Septula of testis
Lobules of testis
Mediastinum testis

Vasculature and Innervation

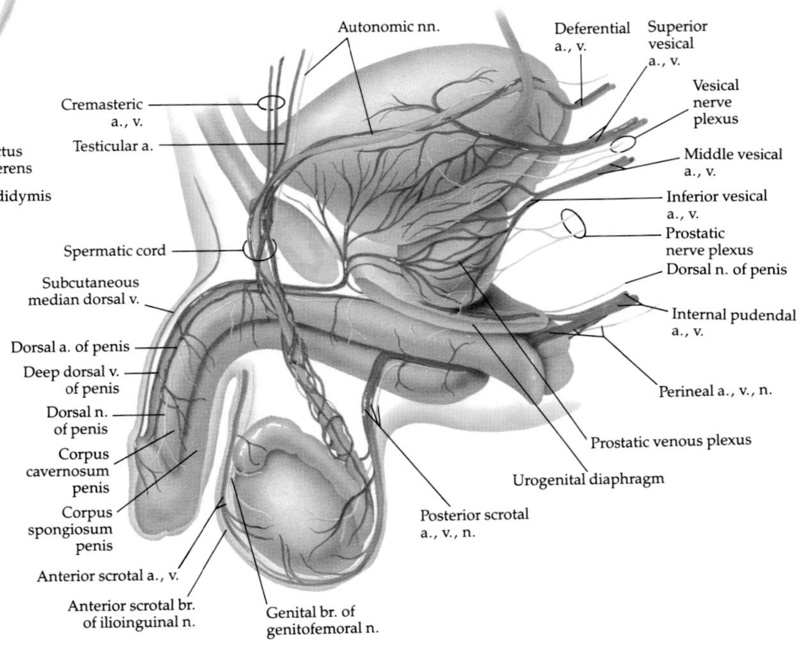

Autonomic nn.
Deferential a., v.
Superior vesical a., v.
Cremasteric a., v.
Vesical nerve plexus
Testicular a.
Middle vesical a., v.
Inferior vesical a., v.
Spermatic cord
Prostatic nerve plexus
Dorsal n. of penis
Subcutaneous median dorsal v.
Internal pudendal a., v.
Dorsal a. of penis
Deep dorsal v. of penis
Perineal a., v., n.
Dorsal n. of penis
Prostatic venous plexus
Corpus cavernosum penis
Urogenital diaphragm
Corpus spongiosum penis
Posterior scrotal a., v., n.
Anterior scrotal a., v.
Anterior scrotal br. of ilioinguinal n.
Genital br. of genitofemoral n.

THE ENDOCRINE SYSTEM

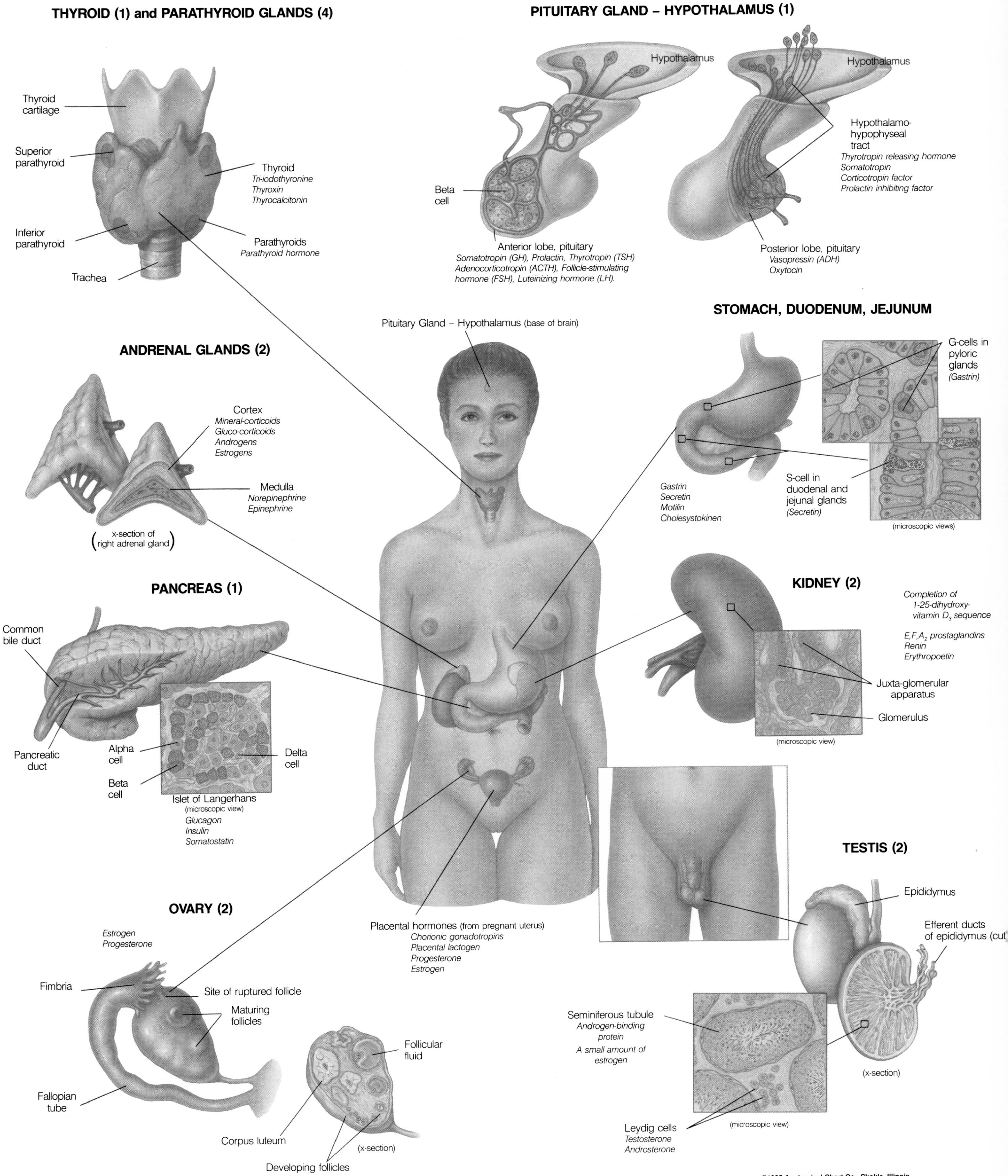
THYROID (1) and PARATHYROID GLANDS (4)
Thyroid cartilage
Superior parathyroid
Thyroid
Tri-iodothyronine
Thyroxin
Thyrocalcitonin
Inferior parathyroid
Parathyroids
Parathyroid hormone
Trachea
PITUITARY GLAND – HYPOTHALAMUS (1)
Hypothalamus
Hypothalamus
Hypothalamo-hypophyseal tract
Thyrotropin releasing hormone
Somatotropin
Corticotropin factor
Prolactin inhibiting factor
Beta cell
Anterior lobe, pituitary
Somatotropin (GH), Prolactin, Thyrotropin (TSH)
Adenocorticotropin (ACTH), Follicle-stimulating hormone (FSH), Luteinizing hormone (LH).
Posterior lobe, pituitary
Vasopressin (ADH)
Oxytocin
Pituitary Gland – Hypothalamus (base of brain)
ANDRENAL GLANDS (2)
Cortex
Mineral-corticoids
Gluco-corticoids
Androgens
Estrogens
Medulla
Norepinephrine
Epinephrine
(x-section of right adrenal gland)
STOMACH, DUODENUM, JEJUNUM
G-cells in pyloric glands
(Gastrin)
Gastrin
Secretin
Motilin
Cholesystokinen
S-cell in duodenal and jejunal glands
(Secretin)
(microscopic views)
PANCREAS (1)
Common bile duct
Pancreatic duct
Alpha cell
Beta cell
Delta cell
Islet of Langerhans
(microscopic view)
Glucagon
Insulin
Somatostatin
KIDNEY (2)
Completion of 1-25-dihydroxy-vitamin D_3 sequence
E,F,A_2 prostaglandins
Renin
Erythropoetin
Juxta-glomerular apparatus
Glomerulus
(microscopic view)
TESTIS (2)
Epididymus
Efferent ducts of epididymus (cut)
Seminiferous tubule
Androgen-binding protein
A small amount of estrogen
(x-section)
Leydig cells
Testosterone
Androsterone
(microscopic view)
OVARY (2)
Estrogen
Progesterone
Fimbria
Site of ruptured follicle
Maturing follicles
Fallopian tube
Follicular fluid
Corpus luteum
Developing follicles
(x-section)
Placental hormones (from pregnant uterus)
Chorionic gonadotropins
Placental lactogen
Progesterone
Estrogen

THE VASCULAR SYSTEM AND VISCERA

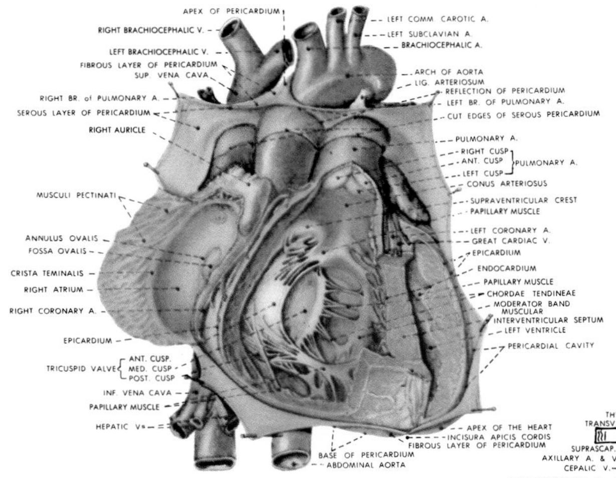

THE INTERIOR OF THE RIGHT ATRIUM AND RIGHT VENTRICLE

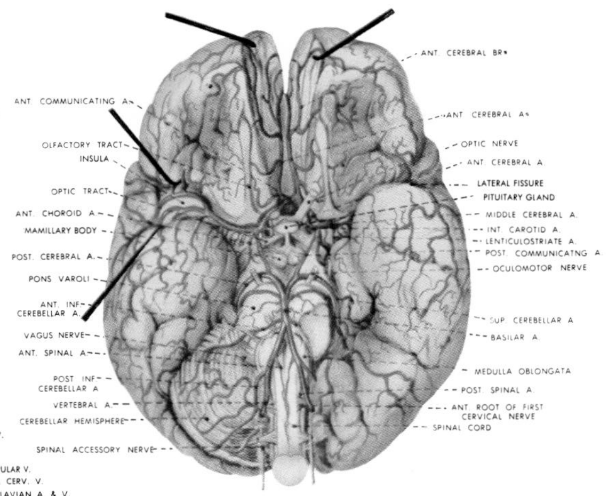

THE ARTERIES OF THE BASE OF THE BRAIN
THE LEFT CEREBELLAR HEMISPHERE HAS BEEN REMOVED

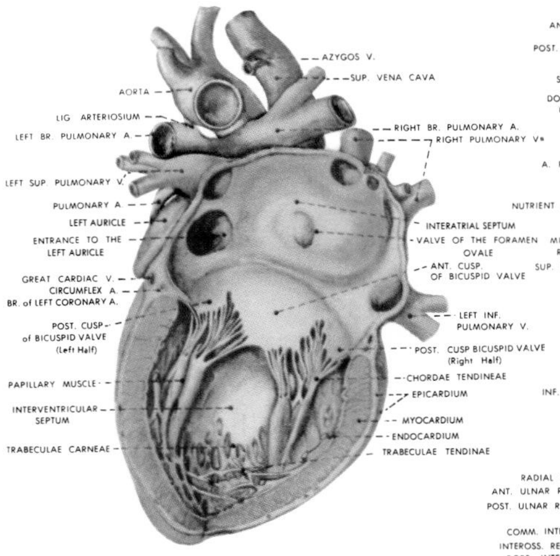

THE INTERIOR OF THE LEFT ATRIUM AND VENTRICLE
SEEN FROM THE LEFT SIDE

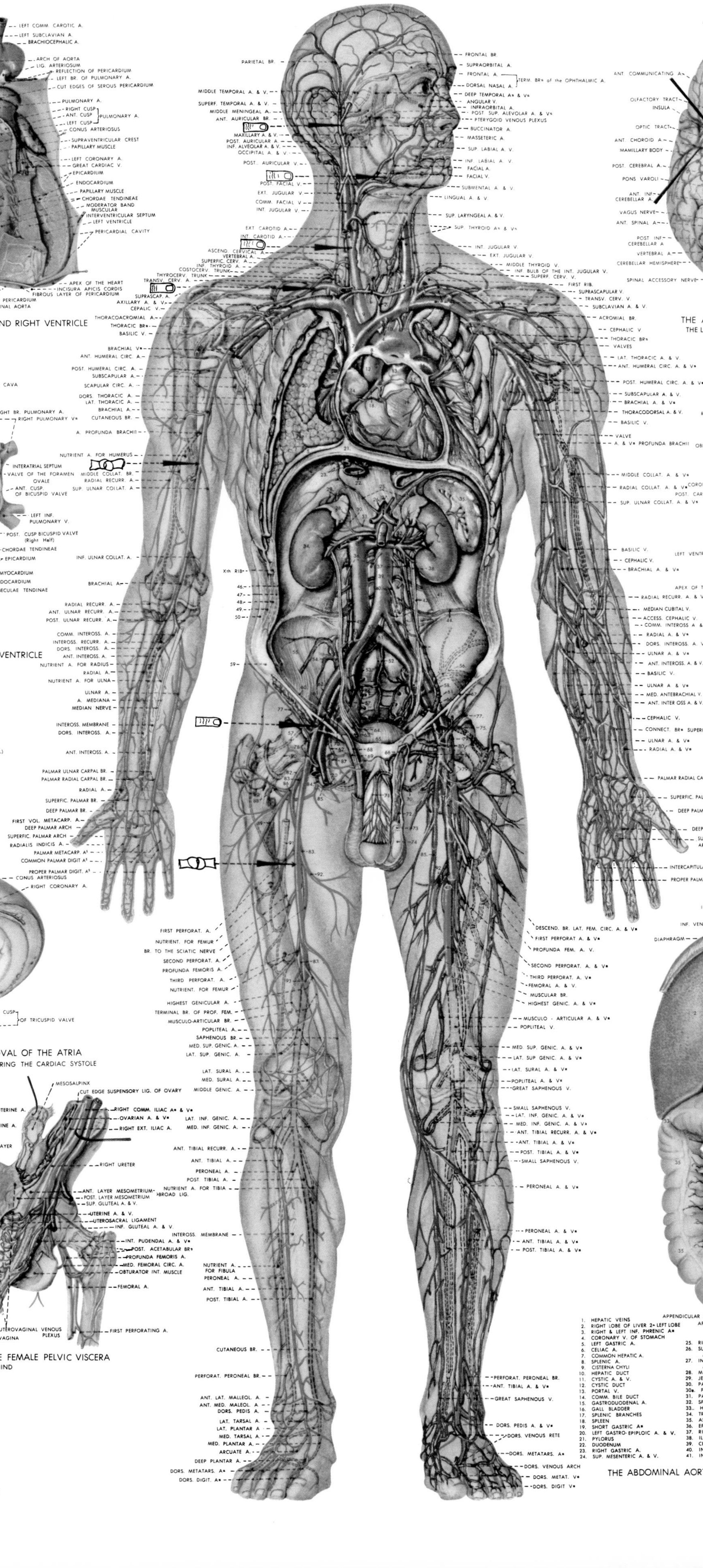

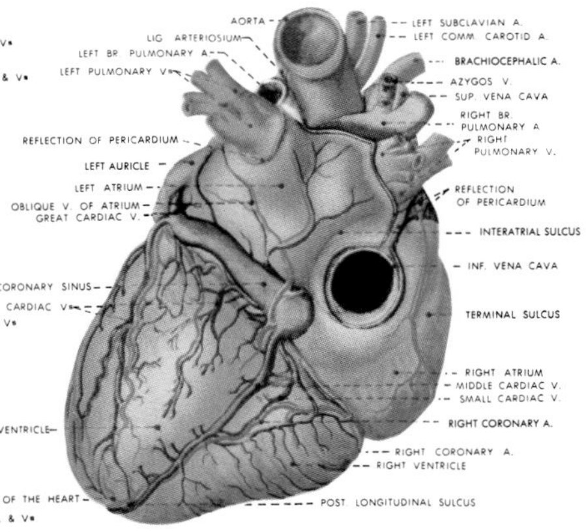

THE ARTERIES AND VEINS OF THE HEART
SEEN FROM BEHIND

POINTS OF APPLICATION FOR TOURNIQUETS

DIGITAL PRESSURE POINTS

CERVICAL REGION

1. JUGULAR VENOUS ARCH
2. THYROIDEA IMA V.
3. RIGHT INT. THORACIC A. & V.

THORACIC REGION

4. THYMIC V▪
5. PARIETAL PLEURA
6. RIGHT BRACHIOCEPHALIC V.
7. SUP. VENA CAVA
8. CUT EDGE OF PERICARDIUM
9. BRACHIOCEPHALIC A.
10 LEFT COMM. CAROTID A.
11. LEFT SUBCLAVIAN A.
12. AORTA
13. PULMONARY A.
14. RIGHT ATRIUM & AURICLE
15. LEFT AURICLE
16. LEFT BR. OF THE PULMONARY A.
17. LEFT PULMONARY V▪
18. RIGHT CORONARY A.
19. ANT. INTERVENTRICULAR A.
20. UPPER LOBE } RIGHT LUNG
20a. MIDDLE LOBE } RIGHT LUNG
20b. LOWER LOBE } RIGHT LUNG

ABDOMINAL REGION

21. DIAPHRAGM
22. INF. VENA CAVA
23. HEPATIC V▪
24. ESOPHAGUS
25. RIGHT & LEFT INF. PHRENIC A▪
26. AORTIC HIATUS
27. CELIAC A.
28. RIGHT SUPRARENAL GLAND
29. SUP. SUPRARENAL BR▪ (Inf. Phrenic A.)
30. INF. SUPRARENAL A.
31. MIDDLE SUPRARENAL A.
32. LEFT RENAL A. & V.
33. SUP. MESENTERIC A.
34. RIGHT KIDNEY
35 SPLEEN
36. RIGHT TESTICULAR A. & V.
37. LEFT TESTICULAR A. & V.
38. URETERIC BR. OF THE RENAL A
39. ABDOMINAL AORTA
40. INF. MESENTERIC A.
41. QUADRATUS LUMBORUM M.
42. LEFT URETER
43. RIGHT & LEFT URETERIC BR▪ OF COMM. ILIAC A.
44. THIRD LUMBAR A. & V.

ABDOMINAL REGION

45. RIGHT ASCENDING LUMBAR V.
46. TRANSVERSALIS FASCIA
47. TRANSVERSUS ABDOMINIS M.
48. INT. ABDOMINAL OBLIQUE M.
49. EXT. ABDOMINAL OBLIQUE M.
50. SKIN

PELVIC REGION

51. RIGHT & LEFT COMM. ILIAC A▪ & V▪
52. MIDDLE SACRAL A. & V.
53. RIGHT INT. ILIAC A. & V.
54. ILIOLUMBAR A. & V.
55. ILIACUS M.
56. SUP. GLUTEAL A. & V▪
57. INF. GLUTEAL A.
58. LEFT EXT. ILIAC A. & V.
59. ANT. SUP. ILIAC SPINE
60. DEEP ILIAC CIRCUMFLEX A.
61. INF. EPIGASTRIC A. & V.
62. OBTURATOR A.
63. LAT. UMBILICAL LIGAMENT
64. SUP. VESICAL A.
65. URINARY BLADDER
66. VAS DEFERENS
67. INGUINAL LIGAMENT
68. CREMASTERIC A.
69 SPERMATIC CORD
70. INT. PUDENDAL A. & V▪
71. DORSAL A▪ & V. OF THE PENIS

LOWER LIMB

72. PAMPINIFORM PLEXUS
73. EPIDIDYMIS
74. TESTICLE
75. FEMORAL A. & V.
76. SUPERF. EPIGASTRIC A.
77. SUPERF. ILIAC CIRC. A.
78. FALCIFORM EDGE of the FOSSA OVALIS
79. VASA VASORUM
80. SUPERF. EXT. PUDENDAL A. & V▪
81. DEEP EXT. PUDENDAL A. & V▪
82. A. COMITANS NERVI ISCHIADICI
83. PROFUNDA FEMORIS A.
84. MED. FEMORAL CIRC. A.
85. GREAT SAPHENOUS V.
86. LAT. FEMORAL CIRC. A.
87. ARTICULAR BR. of the OBTURATOR A.
88. ASCENDING BR. LAT. FEM. CIRC. A.
89. DESCENDING BR. LAT. FEM. CRC. A.
90. ANT. ACETABULAR BRANCH
91. SCIATIC NERVE
92. MUSCULAR BR.
93. ADDUCTOR HIATUS

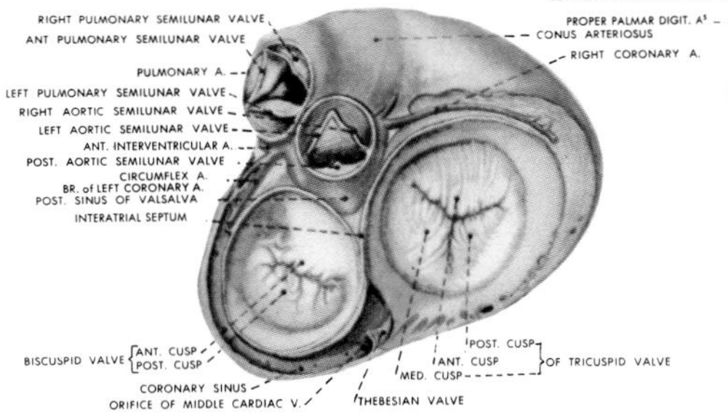

BASE OF VENTRICLES EXPOSED BY REMOVAL OF THE ATRIA
THE VALVES ARE IN THE POSITION THAT OBTAINS DURING THE CARDIAC SYSTOLE

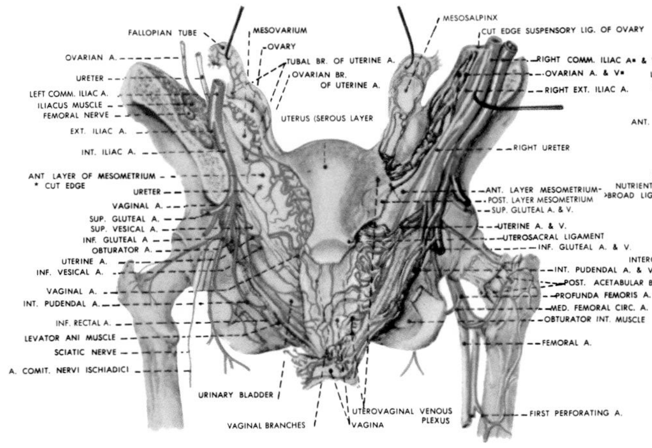

THE ARTERIES AND VEINS OF THE FEMALE PELVIC VISCERA
SEEN FROM BEHIND

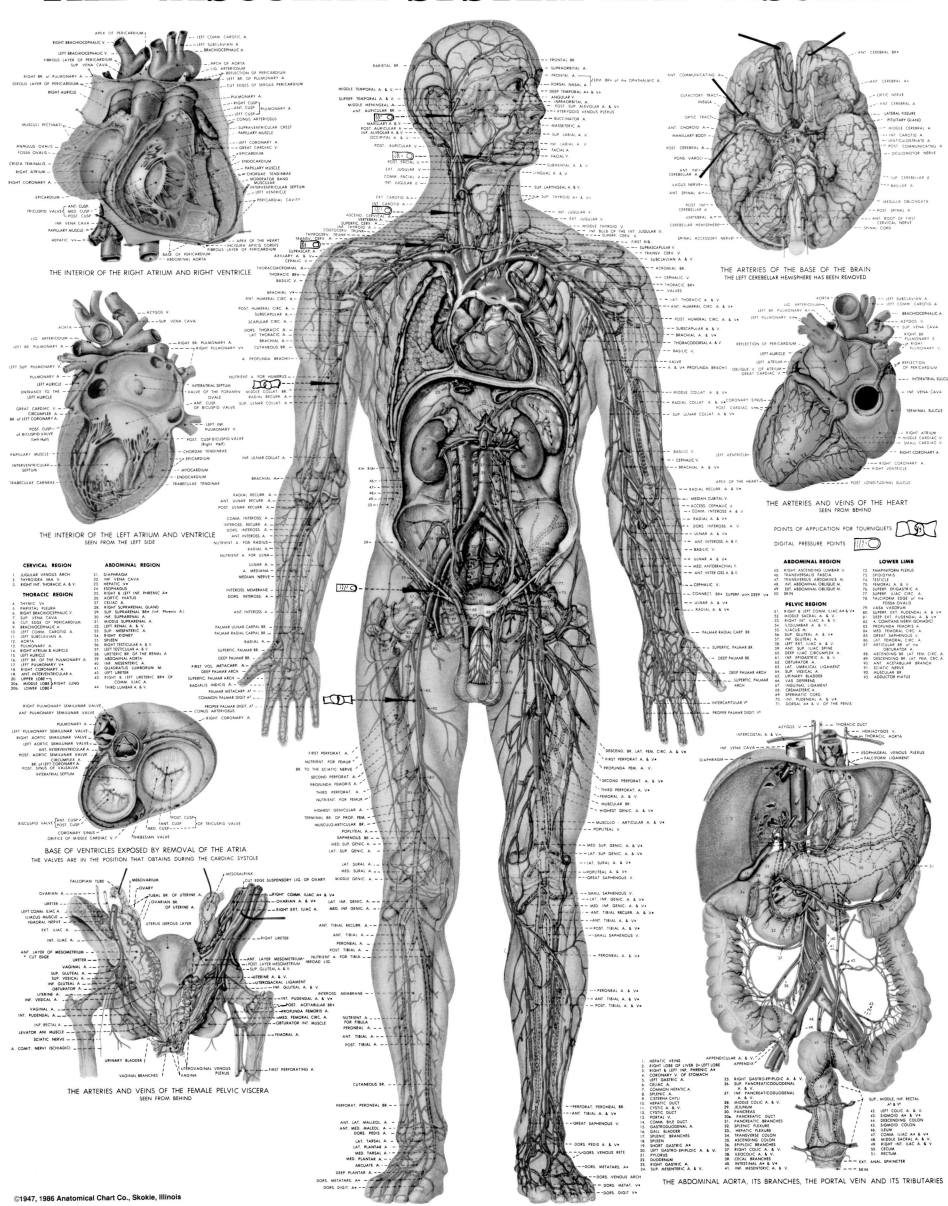

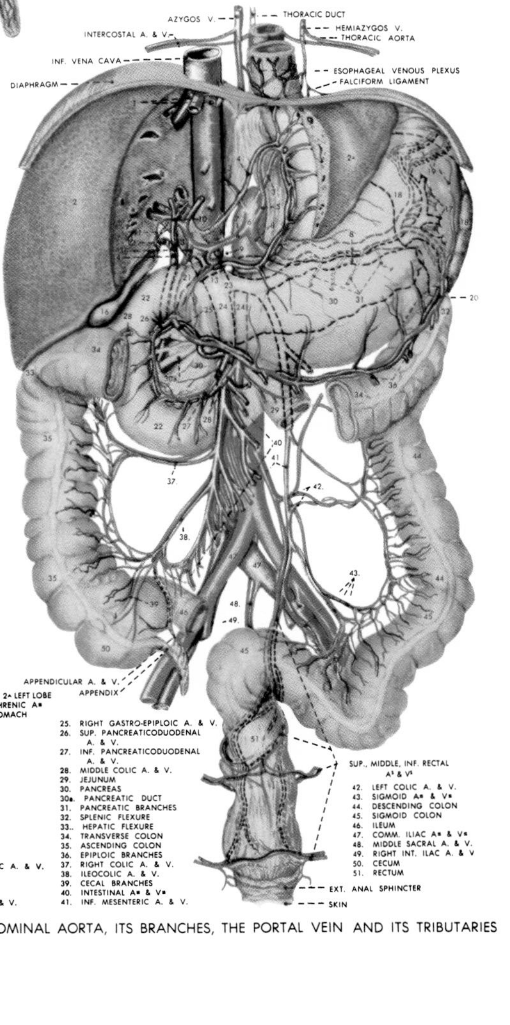

1. HEPATIC VEINS
2. RIGHT LOBE OF LIVER 2a LEFT LOBE
3. RIGHT & LEFT INF. PHRENIC A▪
4. CORONARY V. OF STOMACH
5. LEFT GASTRIC A.
6. CELIAC A.
7. COMMON HEPATIC A.
8. SPLENIC A.
9. CISTERNA CHYLI
10. HEPATIC DUCT
11. CYSTIC A. & V.
12. CYSTIC DUCT
13. PORTAL V.
14. COMM. BILE DUCT
15. GASTRODUODENAL A.
16. GALL BLADDER
17. SPLENIC BRANCHES
18. SPLEEN
19. SHORT GASTRIC A▪
20. LEFT GASTRO-EPIPLOIC A. & V.
21. PYLORUS
22. DUODENUM
23. RIGHT GASTRIC A.
24. SUP. MESENTERIC A. & V.
25. RIGHT GASTRO-EPIPLOIC A. & V.
26. SUP. PANCREATICODUODENAL A. & V.
27. INF. PANCREATICODUODENAL A. & V.
28. MIDDLE COLIC A. & V.
29. JEJUNUM
30. PANCREAS
30a. PANCREATIC DUCT
31. PANCREATIC BRANCHES
32. SPLENIC FLEXURE
33.. HEPATIC FLEXURE
34. TRANSVERSE COLON
35. ASCENDING COLON
36. EPIPLOIC BRANCHES
37. RIGHT COLIC A. & V.
38. ILEOCOLIC A. & V.
39. CECAL BRANCHES
40. INTESTINAL A▪ & V▪
41. INF. MESENTERIC A. & V.
42. LEFT COLIC A. & V.
43. SIGMOID A▪ & V▪
44. DESCENDING COLON
45. SIGMOID COLON
46. ILEUM
47. COMM. ILIAC A▪ & V▪
48. MIDDLE SACRAL A. & V.
49. RIGHT INT. ILAC A. & V
50. CECUM
51. RECTUM

THE ABDOMINAL AORTA, ITS BRANCHES, THE PORTAL VEIN AND ITS TRIBUTARIES

THE HEART

ANTERIOR VIEW OF HEART AND GREAT VESSELS

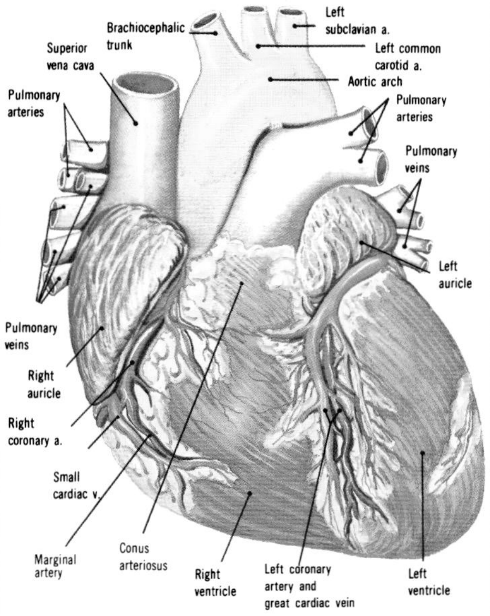

CORONARY BLOOD VESSELS

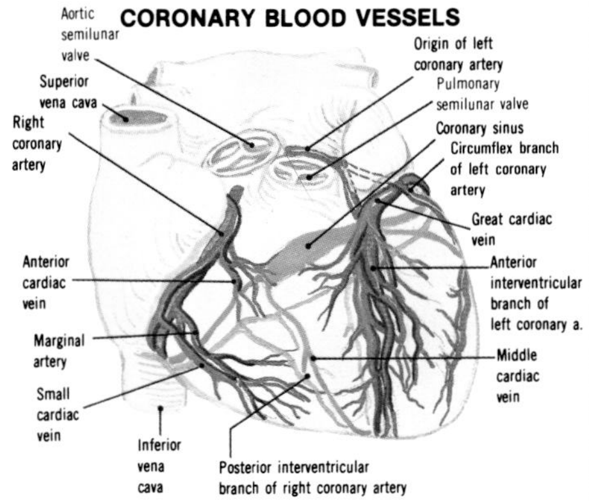

SECTION OF HEART WALL

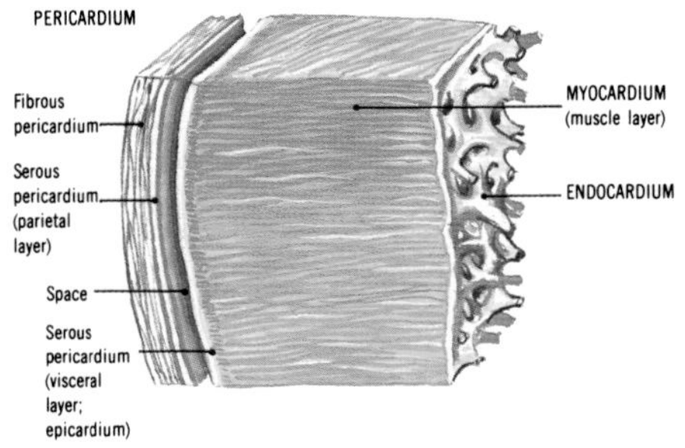

CARDIAC MUSCLE TISSUE

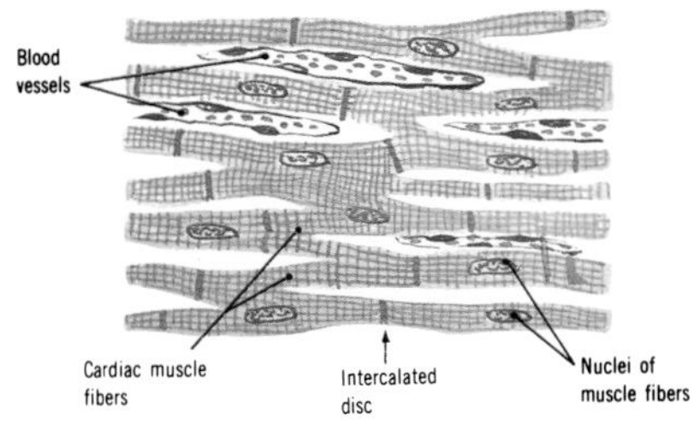

CONDUCTING SYSTEM

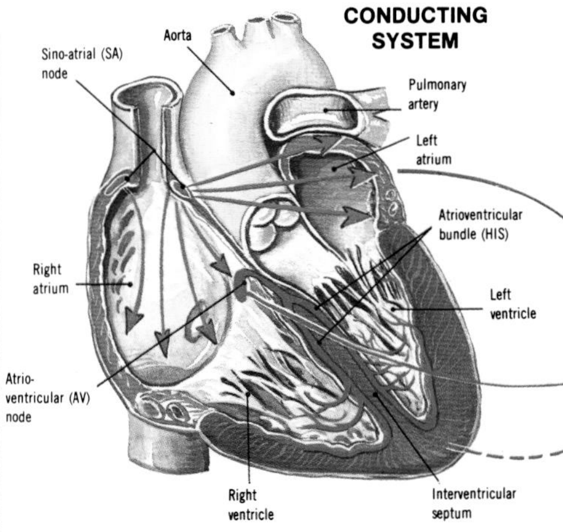

BLOOD FLOW THROUGH THE HEART

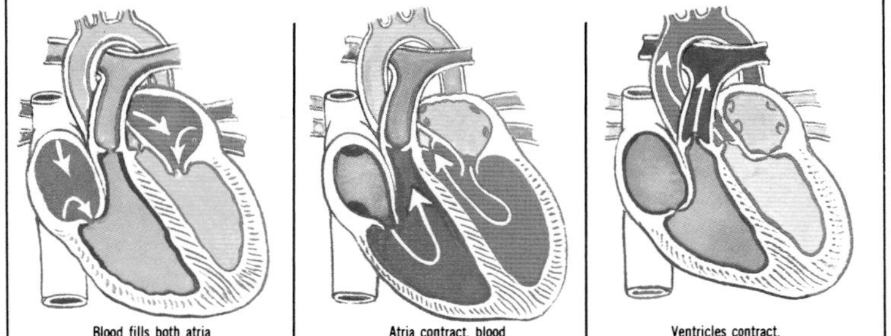

Blood fills both atria

Atria contract, blood enters ventricles

Ventricles contract, blood enters into aorta and pulmonary arteries

HEART CHAMBERS, VALVES AND VESSELS

CHAMBERS:
1. Right atrium
2. Left atrium
3. Right ventricle
4. Left ventricle

VESSELS:
5. Ascending aorta
6. Aortic arch
7. Descending aorta
8. Brachiocephalic trunk
9. Left common carotid a.
10. Left subclavian a.
11. Pulmonary trunk
12. Pulmonary arteries, right
13. Pulmonary arteries, left
14. Pulmonary veins, right
15. Pulmonary veins, left
16. Superior vena cava
17. Inferior vena cava
18. Right coronary artery & small cardiac vein
19. Left coronary artery & great cardiac vein

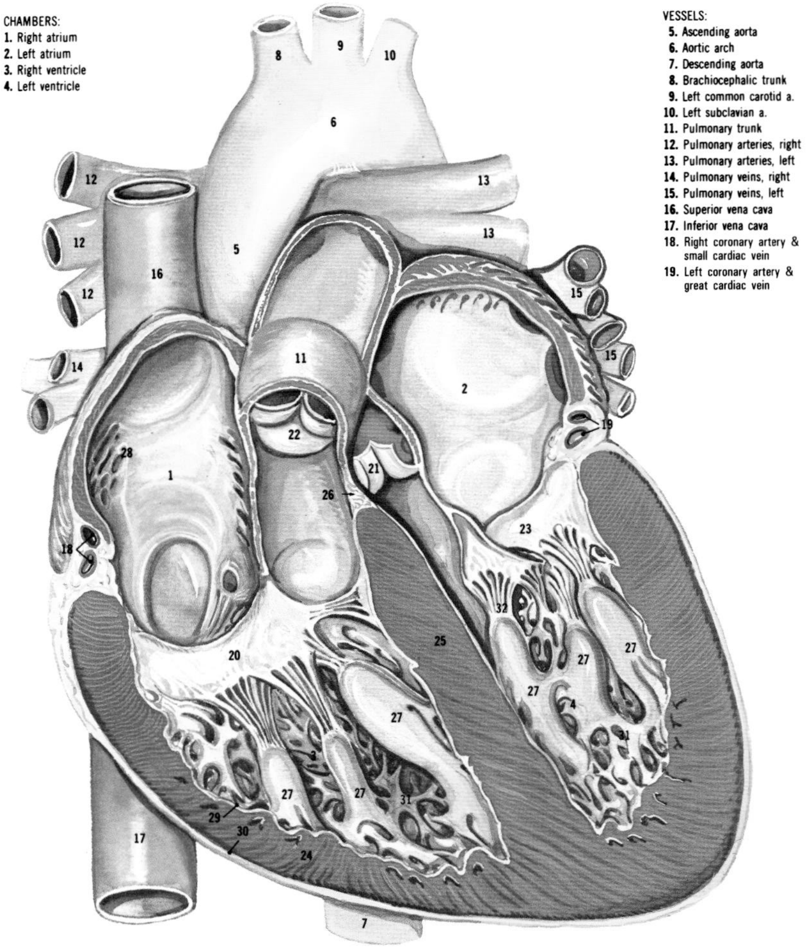

VALVES:
20. Tricuspid
21. Aortic (semilunar)
22. Pulmonary (semilunar)
23. Mitral (biscupid)

HEART WALL AND MUSCLES
24. Myocardium
25. Interventricular septum (muscle portion)
26. Interventricular septum (membranous portion)
27. Papillary muscles
28. Pectinate muscles
29. Endocardium
30. Epicardium
31. Trabeculae carnae

CORDS:
32. Chordae tendineae
33. Interventricular

POSTERIOR VIEW OF HEART AND GREAT VESSELS

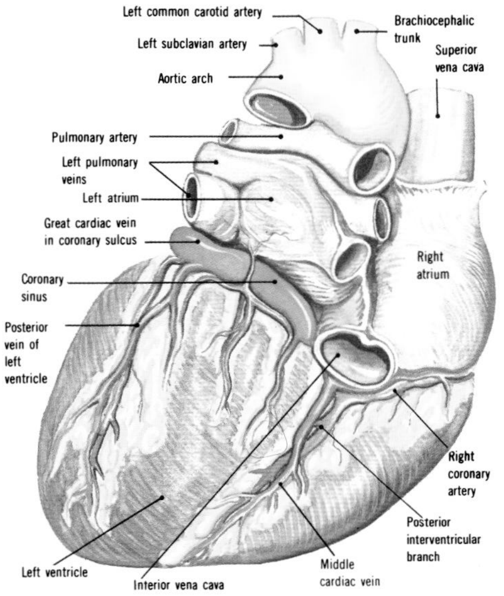

VALVES OF HEART VIEWED FROM ABOVE (ATRIA REMOVED)

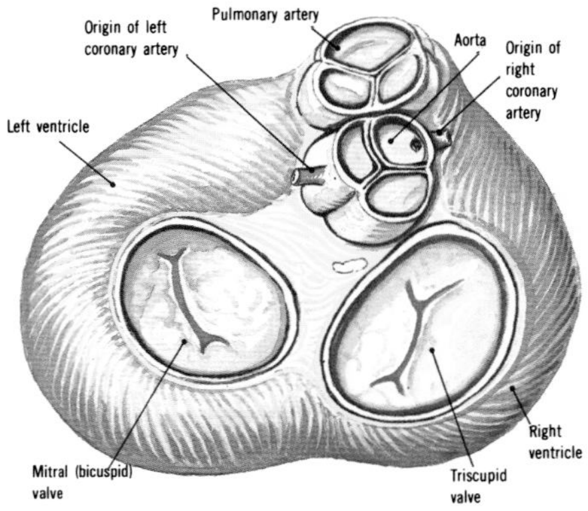

PROJECTION OF HEART VALVES AND AUSCULTATION SITES ON ANTERIOR THORAX

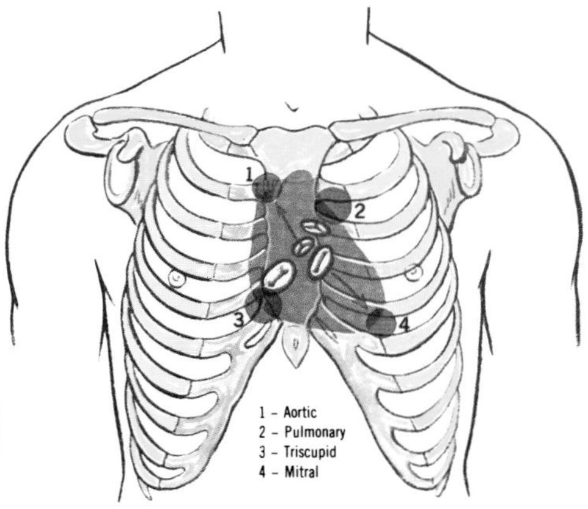

ELECTROCARDIOGRAM

P = depolarization process throughout the atria

QRS = depolarization process throughout the ventricles

T = repolarization of ventricles

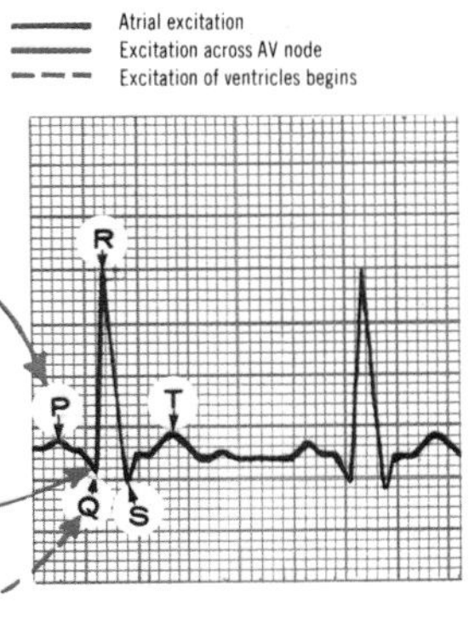

CORONARY OCCLUSION

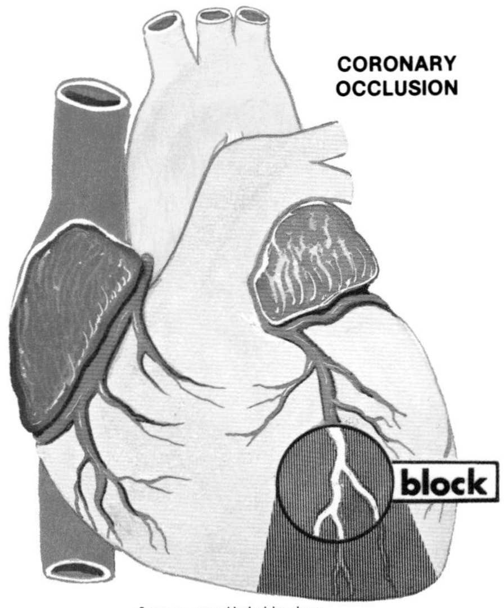

Coronary artery blocked by plaque. Heart muscle beyond occlusion is deprived of blood causing ventricle to contract incompletely.

EXTERNAL CARDIAC COMPRESSION

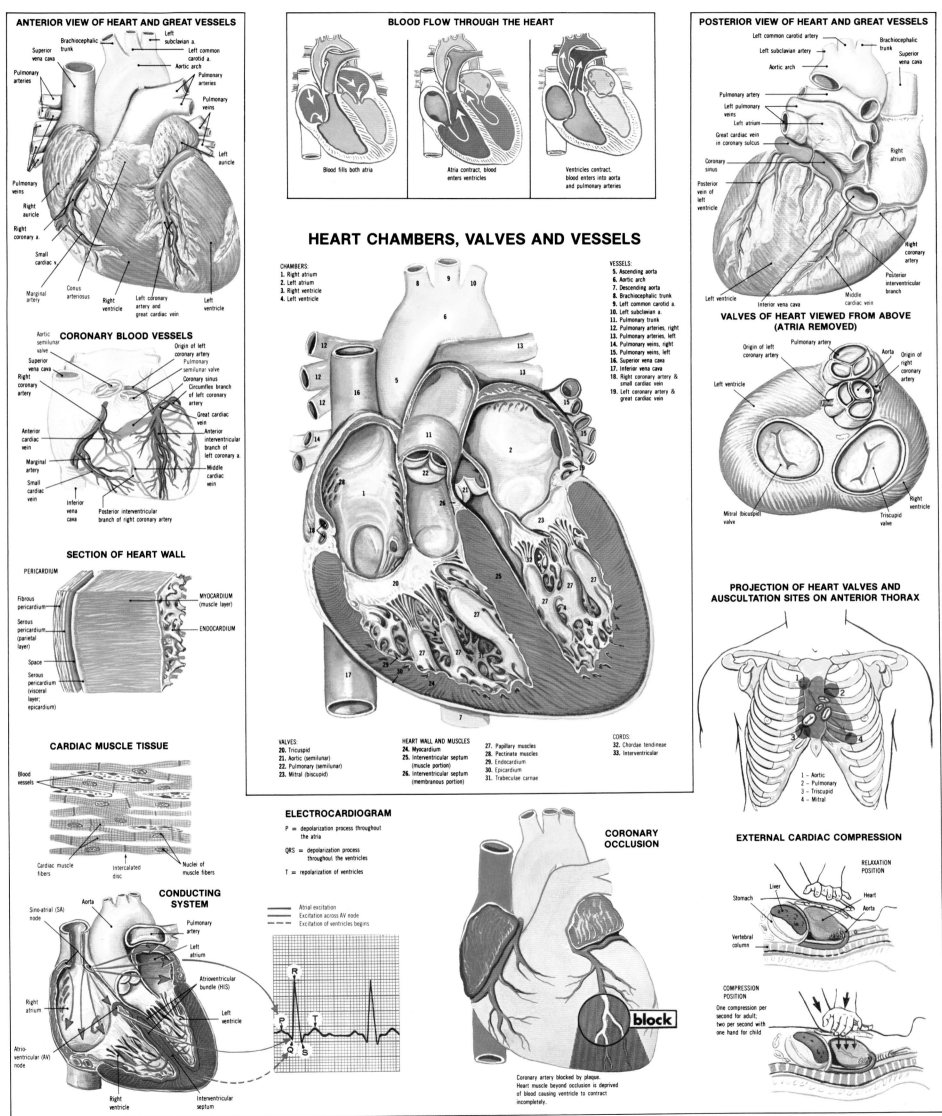

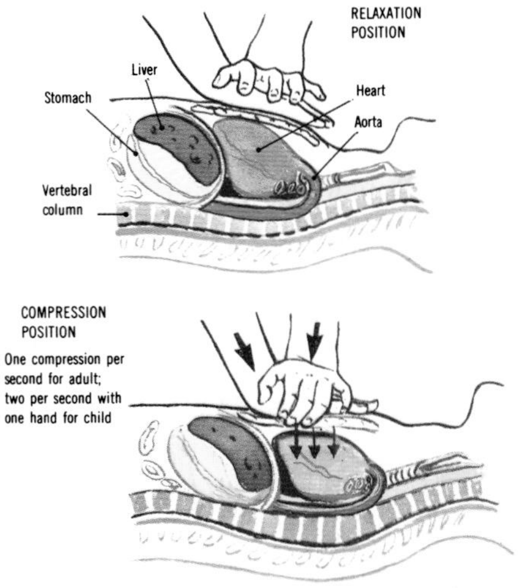

THE BRAIN

ARTERIES OF THE BRAIN — lateral view

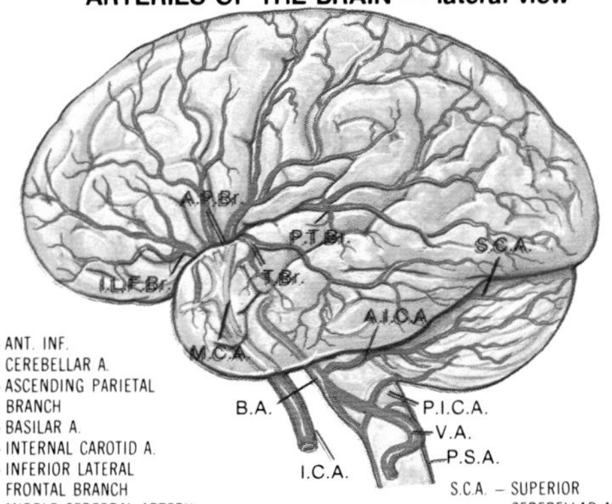

A.I.C.A. – ANT. INF. CEREBELLAR A.
A.P.Br. – ASCENDING PARIETAL BRANCH
B.A. – BASILAR A.
I.C.A. – INTERNAL CAROTID A.
I.L.F.Br. – INFERIOR LATERAL FRONTAL BRANCH
M.C.A. – MIDDLE CEREBRAL ARTERY
P.I.C.A. – POST. INF. CEREBELLAR A.
P.S.A. – POSTERIOR SPINAL A.
P.T.Br. – PARIETO-TEMPORAL BRANCH
S.C.A. – SUPERIOR CEREBELLAR A.
T.Br. – TEMPORAL BRANCH
V.A. – VERTEBRAL A.

ARTERIES OF THE BRAIN — median sagittal section

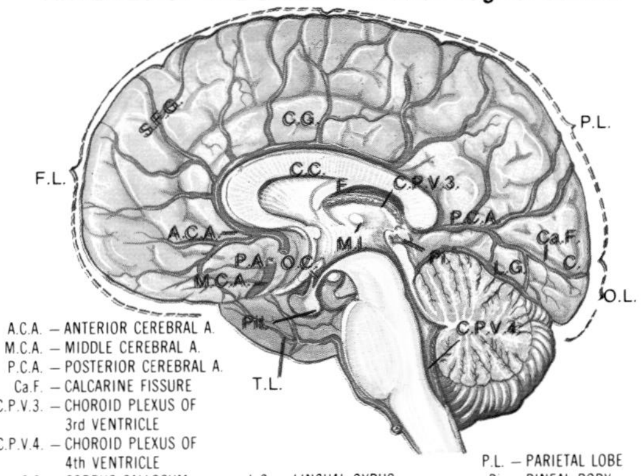

A.C.A. – ANTERIOR CEREBRAL A.
M.C.A. – MIDDLE CEREBRAL A.
P.C.A. – POSTERIOR CEREBRAL A.
Ca.F. – CALCARINE FISSURE
C.P.V.3. – CHOROID PLEXUS OF 3rd VENTRICLE
C.P.V.4. – CHOROID PLEXUS OF 4th VENTRICLE
C.C. – CORPUS CALLOSUM
C.G. – CINGULATE GYRUS
C. – CUNEUS
F. – FORNIX
F.L. – FRONTAL LOBE
L.G. – LINGUAL GYRUS
M.I. – MASSA INTERMEDIA
O.L. – OCCIPITAL LOBE
O.C. – OPTIC CHIASMA
P.A. – PAROLFACTORY AREA
P.L. – PARIETAL LOBE
Pi. – PINEAL BODY
Pit. – PITUITARY GLAND
P.C. – PRECUNEUS
S.F.G. – SUPERIOR FRONTAL GYRUS
T.L. – TEMPORAL LOBE

CORONAL SECTION OF THE BRAIN

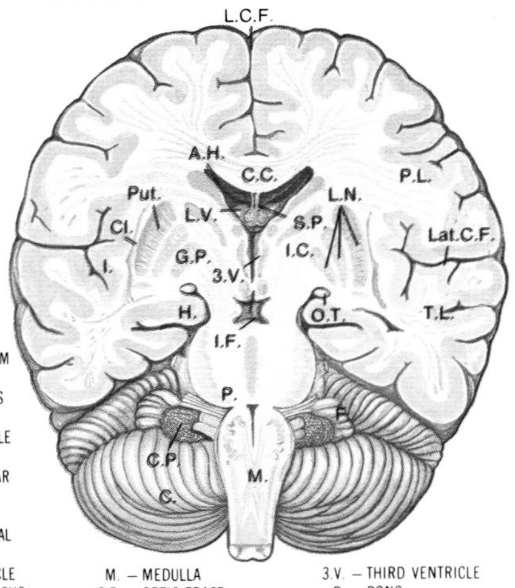

A.H. – ANTERIOR HORN OF LATERAL VENTRICLE
C. – CEREBELLUM
C.P. – CHOROID PLEXUS
Cl. – CLAUSTRUM
C.C. – CORPUS CALLOSUM
F. – FLOCCULUS
G.P. – GLOBUS PALLIDUS
H. – HIPPOCAMPUS
I.C. – INTERNAL CAPSULE
I. – INSULA
I.F. – INTERPEDUNCULAR FOSSA
Lat. C.F. – LATERAL CEREBRAL FISSURE
L.V. – LATERAL VENTRICLE
L.N. – LENTIFORM NUCLEUS
L.C.F. – LONGITUDINAL CEREBRAL FISSURE
M. – MEDULLA
O.T. – OPTIC TRACT
P.L. – PARIETAL LOBE
T.L. – TEMPORAL LOBE
3.V. – THIRD VENTRICLE
P. – PONS
Put. – PUTAMEN
S.P. – SEPTUM PELLUCIDUM

MENINGES OF THE BRAIN

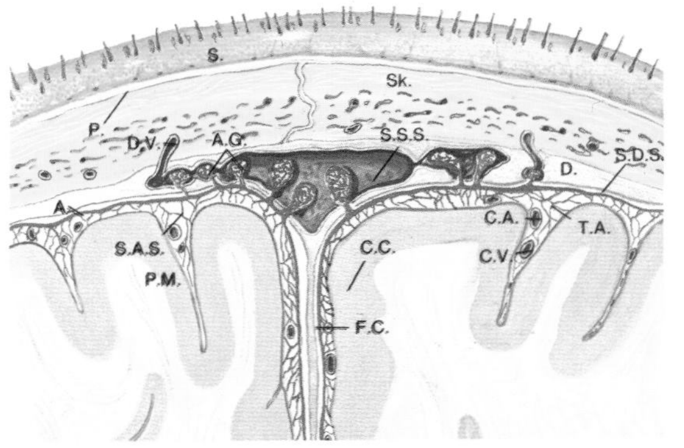

A. – ARACHNOID
A.G. – ARACHNOID GRANULATIONS
C.A. – CEREBRAL ARTERY
C.C. – CEREBRAL CORTEX
C.V. – CEREBRAL VEIN
D.V. – DIPLOIC VEIN
D. – DURA MATER
F.C. – FALX CEREBRI
P. – PERIOSTEUM
P.M. – PIA MATER
S. – SCALP
Sk. – SKULL
S.A.S. – SUBARACHNOID SPACE
S.D.S. – SUBDURAL SPACE
S.S.S. – SUPERIOR SAGITTAL SINUS
T.A. – TRABECULAE OF ARACHNOID

BASE OF BRAIN — ORIGIN OF CRANIAL NERVES

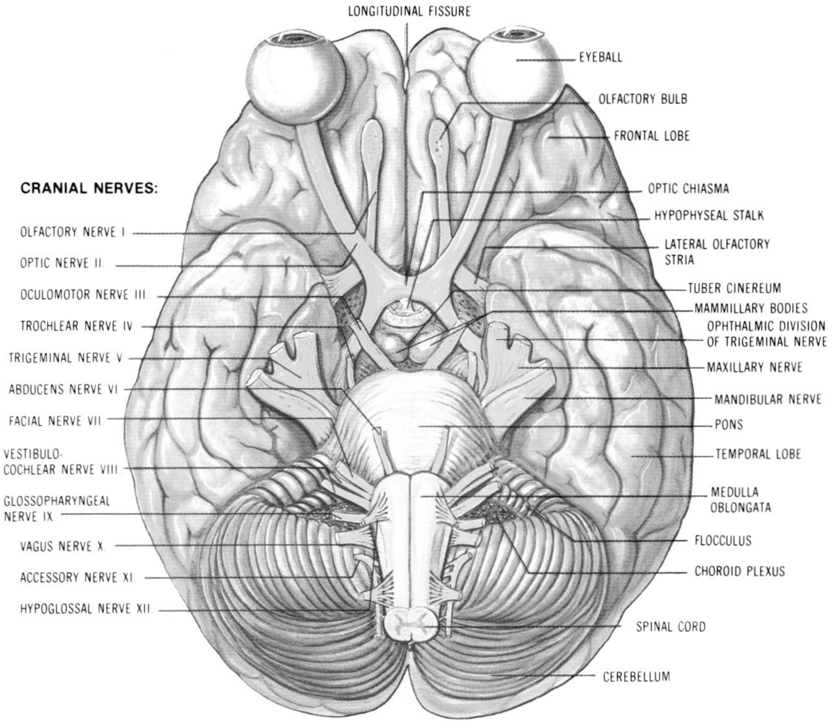

INFERIOR SURFACE OF THE BRAIN WITH BLOOD VESSELS

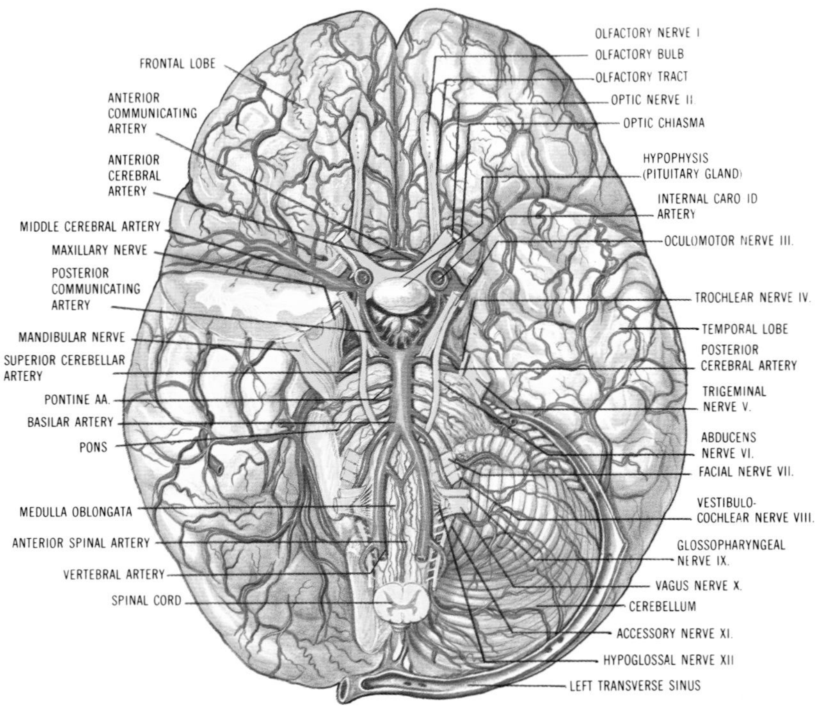

SOMATOTOPIC ORGANIZATION OF THE CEREBRUM

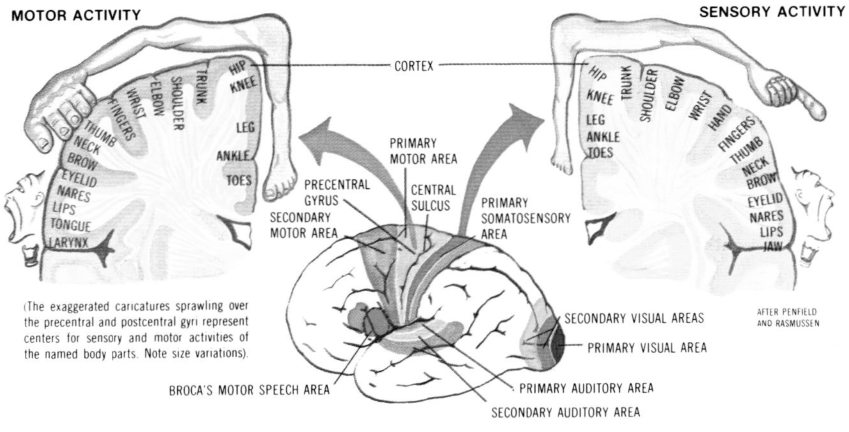

(The exaggerated caricatures sprawling over the precentral and postcentral gyri represent centers for sensory and motor activities of the named body parts. Note size variations).

THE LIMBIC SYSTEM

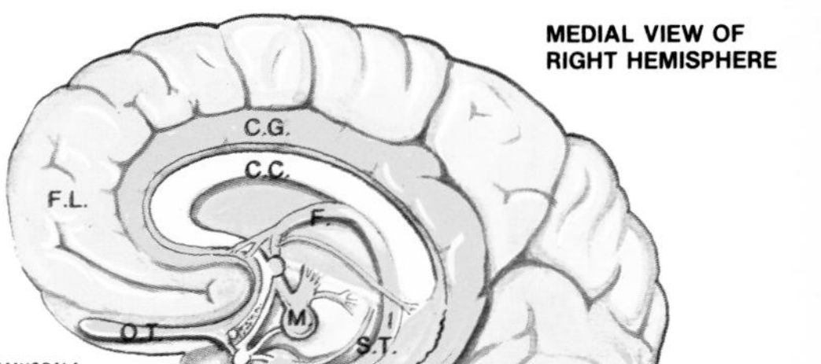

A. – AMYGDALA
C.G. – CINGULATE GYRUS
C.C. – CORPUS CALLOSUM
F. – FORNIX
F.L. – FRONTAL LOBE
H. – HIPPOCAMPUS
M. – MAMMILLARY BODY
O.T. – OLFACTORY TRACT AND BULB
S.T. – STRIA TERMINALIS

The Limbic system includes the thalamus, the hypothalamus, amygdala, parts of the reticular formation in the brain stem, and the limbic region of the cerebral cortex. Its functions deal primarily with behavior and emotions.

VENTRICLES OF THE BRAIN

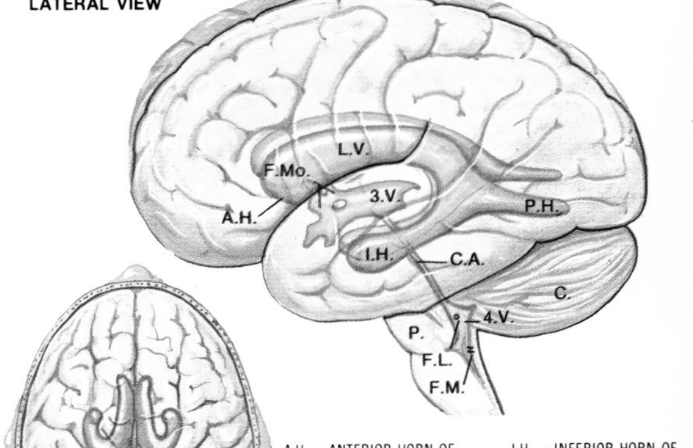

A.H. – ANTERIOR HORN OF LATERAL VENTRICLE
C. – CEREBELLUM
C.A. – CEREBRAL AQUEDUCT
F.L. – FORAMEN OF LUSCHKA
F.M. – FORAMEN OF MAGENDIE
F.Mo. – FORAMEN OF MONRO
4.V. – FOURTH VENTRICLE
I.H. – INFERIOR HORN OF LATERAL VENTRICLE
L.V. – LATERAL VENTRICLE
P. – PONS
P.H. – POSTERIOR HORN OF LATERAL VENTRICLE
3.V. – THIRD VENTRICLE

CIRCLE OF WILLIS

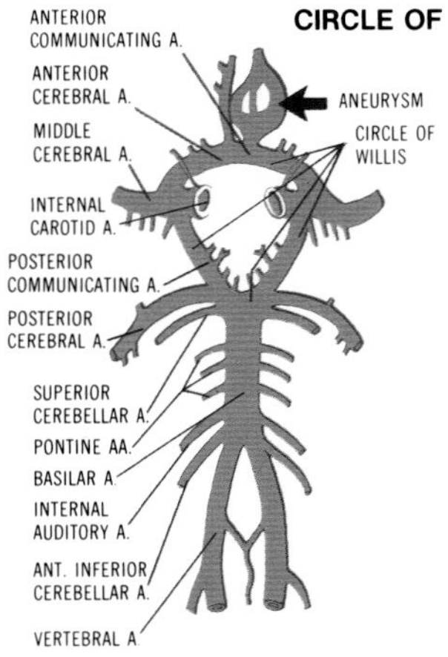

STROKE

An intracranial aneurysm (shown at left) is a vascular dilatation resulting from a localized defect in elasticity of the vessel. An aneurysm may occur anywhere within the circle of Willis at the base of the brain or in any of the branches.

An aneurysm that ruptures is only one possible cause of cerebral vascular accident, also known as CVA, apoplexy, or stroke. Other causes are (1) blockage of an artery due to an arteriosclerotic plaque or thrombus; (2) an embolus or clot can break off from a major clot elsewhere in the arterial system and travel to the brain; and (3) hypertensive apoplexy which is a sudden effusion of blood into the cerebral tissues as a result of vascular hypertension.

CEREBROSPINAL FLUID CIRCULATION

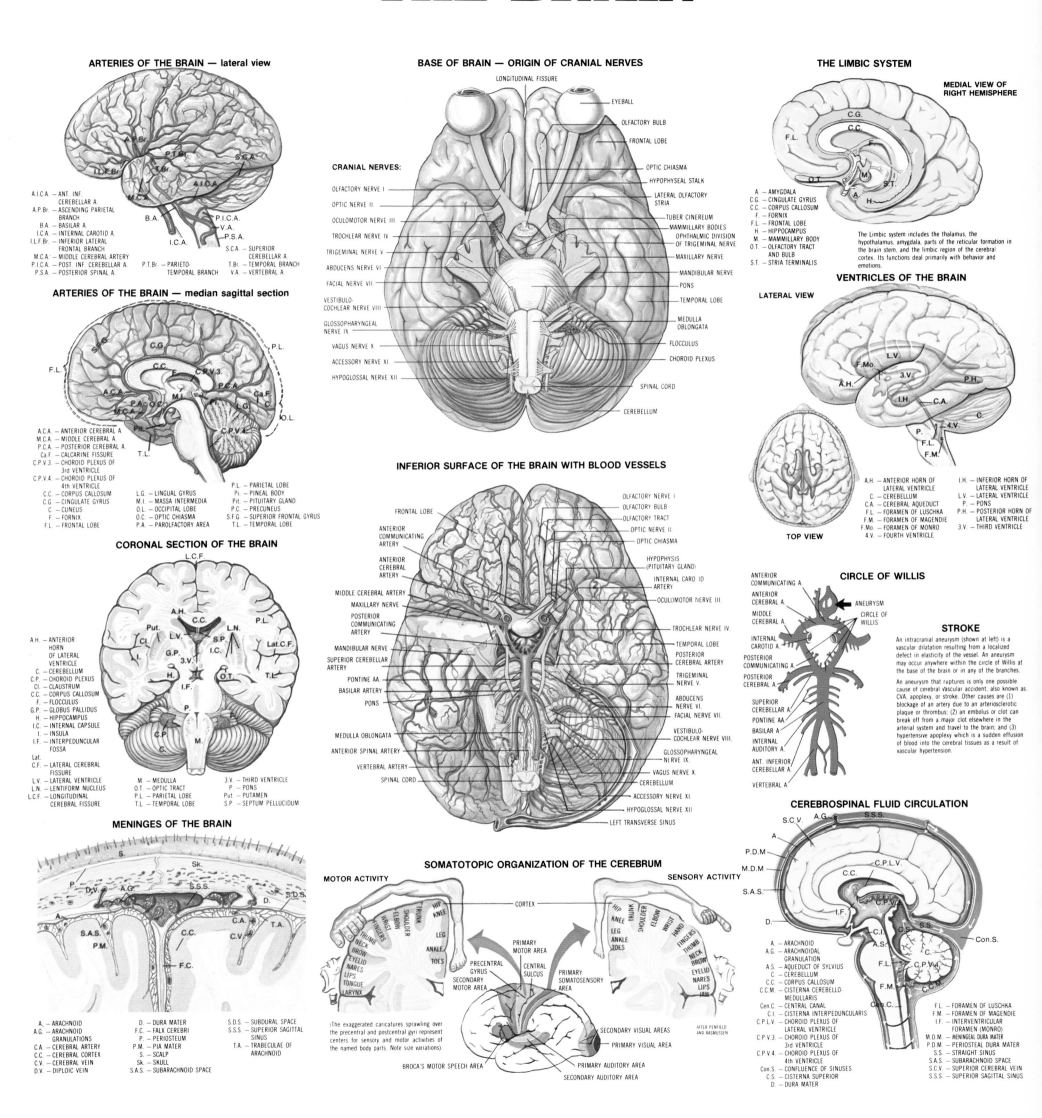

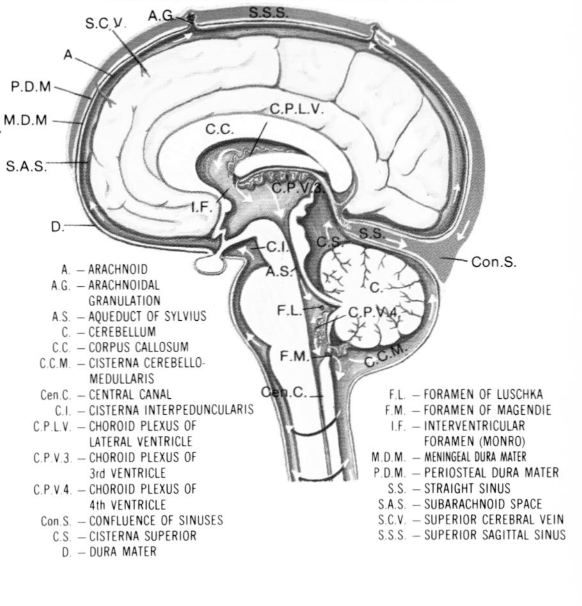

A. – ARACHNOID
A.G. – ARACHNOIDAL GRANULATION
A.S. – AQUEDUCT OF SYLVIUS
C. – CEREBELLUM
C.C. – CORPUS CALLOSUM
C.C.M. – CISTERNA CEREBELLO-MEDULLARIS
Cen.C. – CENTRAL CANAL
C.I. – CISTERNA INTERPEDUNCULARIS
C.P.L.V. – CHOROID PLEXUS OF LATERAL VENTRICLE
C.P.V.3. – CHOROID PLEXUS OF 3rd VENTRICLE
C.P.V.4. – CHOROID PLEXUS OF 4th VENTRICLE
Con.S. – CONFLUENCE OF SINUSES
C.S. – CISTERNA SUPERIOR
D. – DURA MATER
F.L. – FORAMEN OF LUSCHKA
F.M. – FORAMEN OF MAGENDIE
I.F. – INTERVENTRICULAR FORAMEN (MONRO)
M.D.M. – MENINGEAL DURA MATER
P.D.M. – PERIOSTEAL DURA MATER
S.S. – STRAIGHT SINUS
S.A.S. – SUBARACHNOID SPACE
S.C.V. – SUPERIOR CEREBRAL VEIN
S.S.S. – SUPERIOR SAGITTAL SINUS

THE VERTEBRAL COLUMN

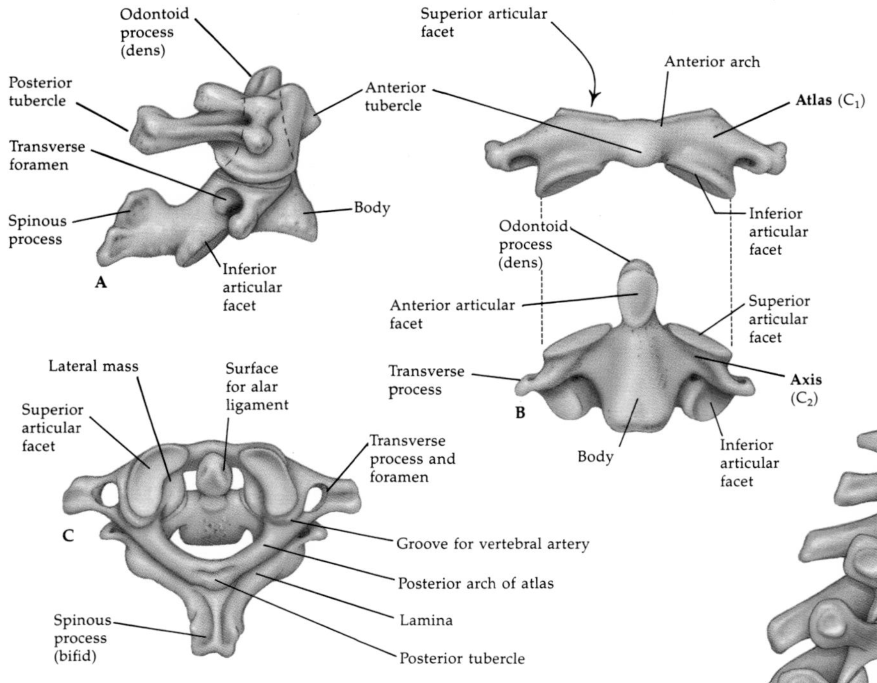

Atlas and Axis
A-Articulated right lateral view, B-Disarticulated anterior view, C-Articulated posterior superior view

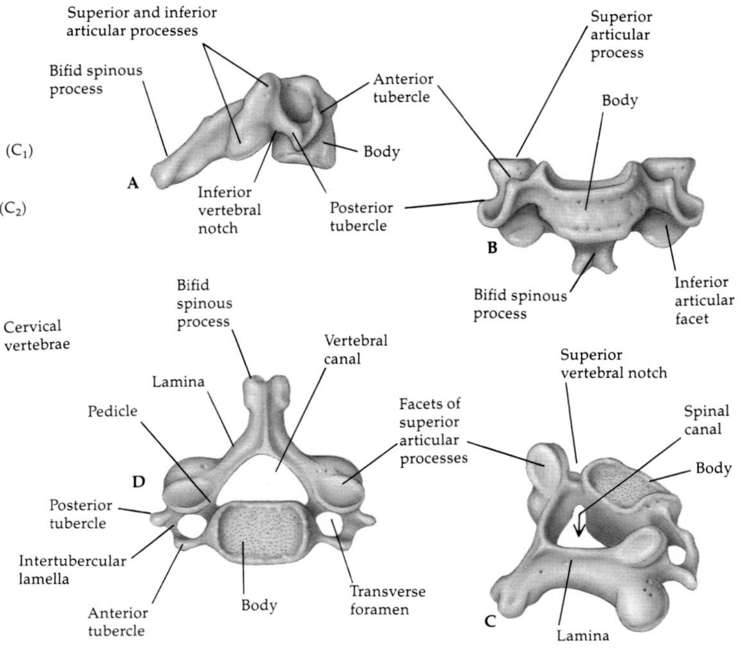

Fifth Cervical Vertebra
A-Lateral view, B-Anterior view, C-Posterolateral oblique view, D-Superior view

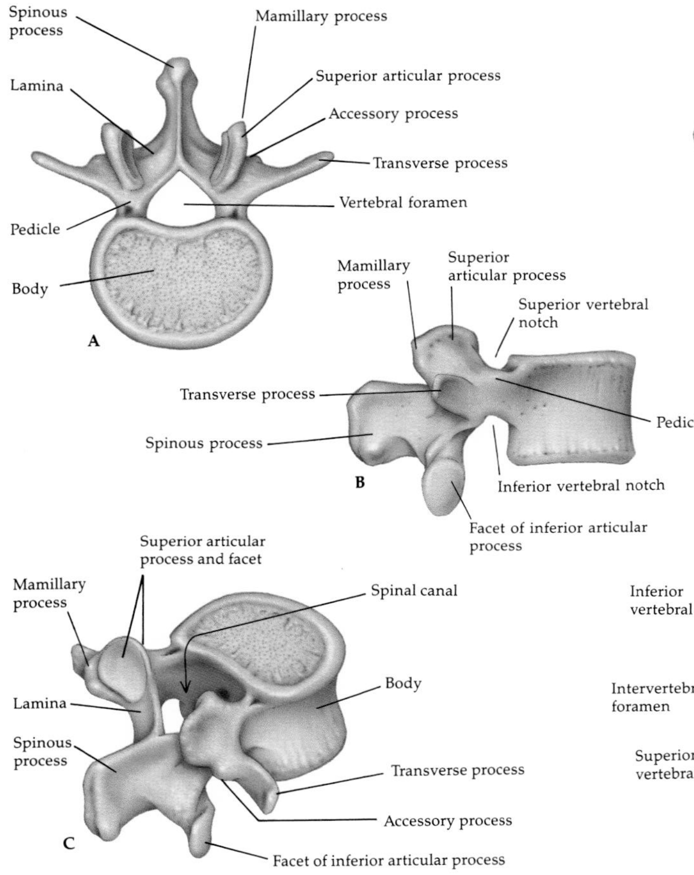

Second Lumbar Vertebra
A-Superior view, B-Lateral view, C-Posterolateral oblique view

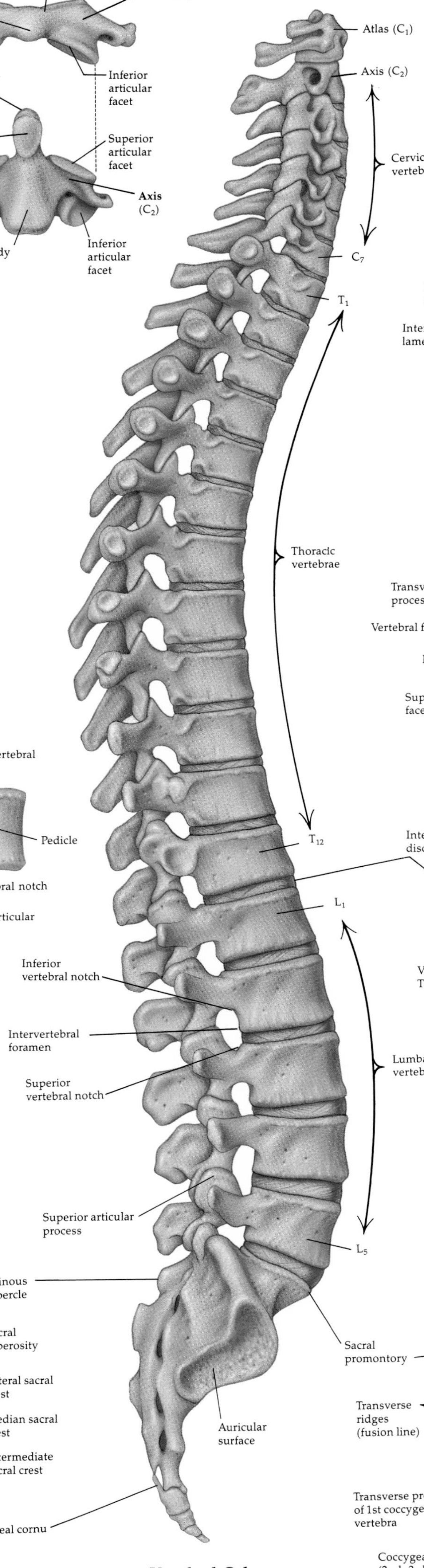

Vertebral Column
(Right Lateral View)

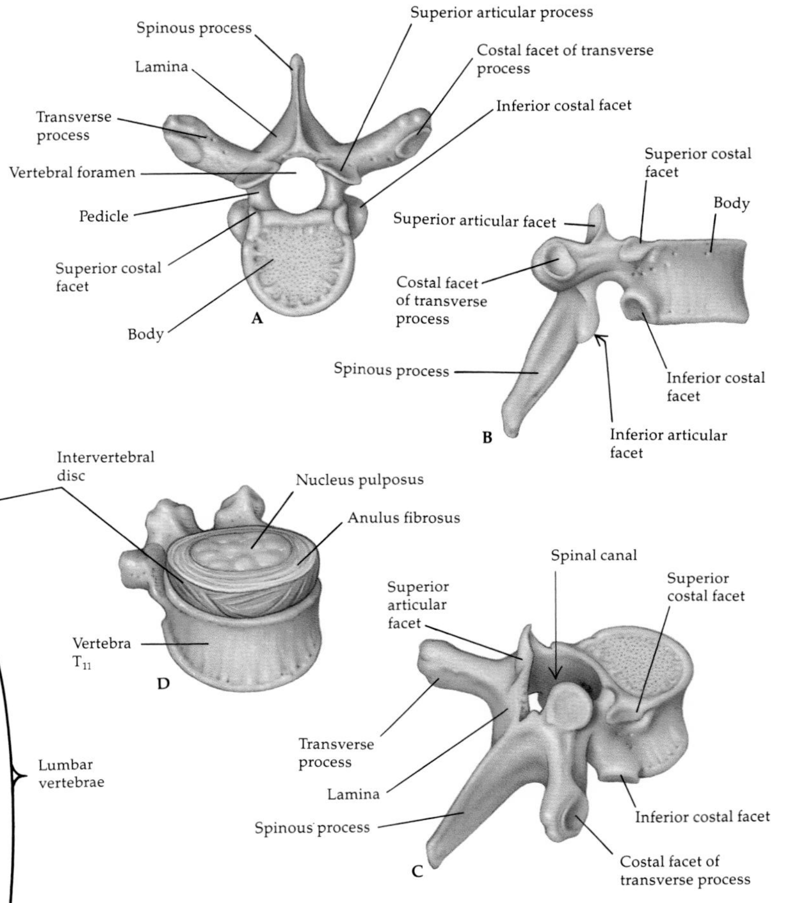

Seventh and Eleventh Thoracic Vertebrae
A-Superior view, B-Lateral view, C-Posterolateral oblique view, D-Anterolateral oblique view of T_{11}

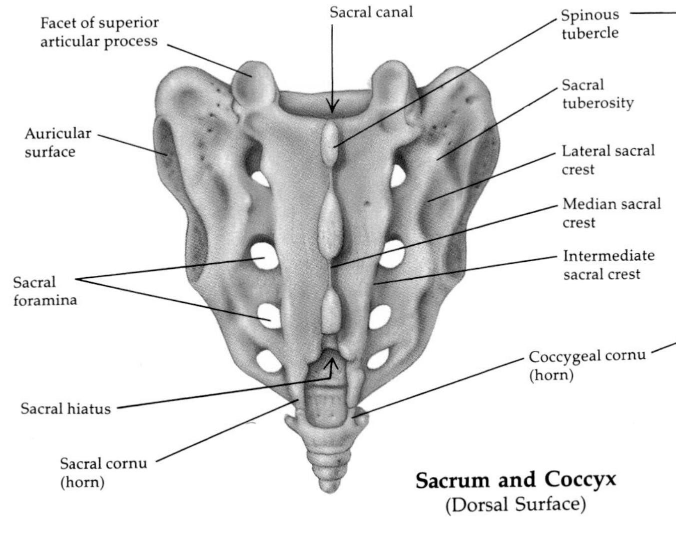

Sacrum and Coccyx
(Dorsal Surface)

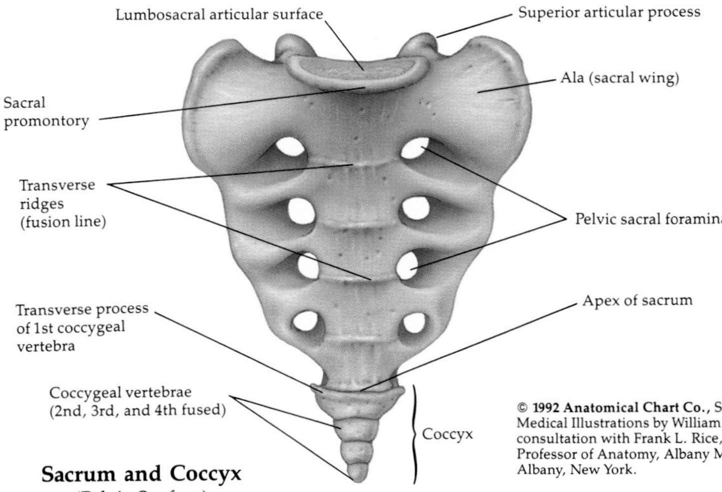

Sacrum and Coccyx
(Pelvic Surface)

THE HUMAN SKULL

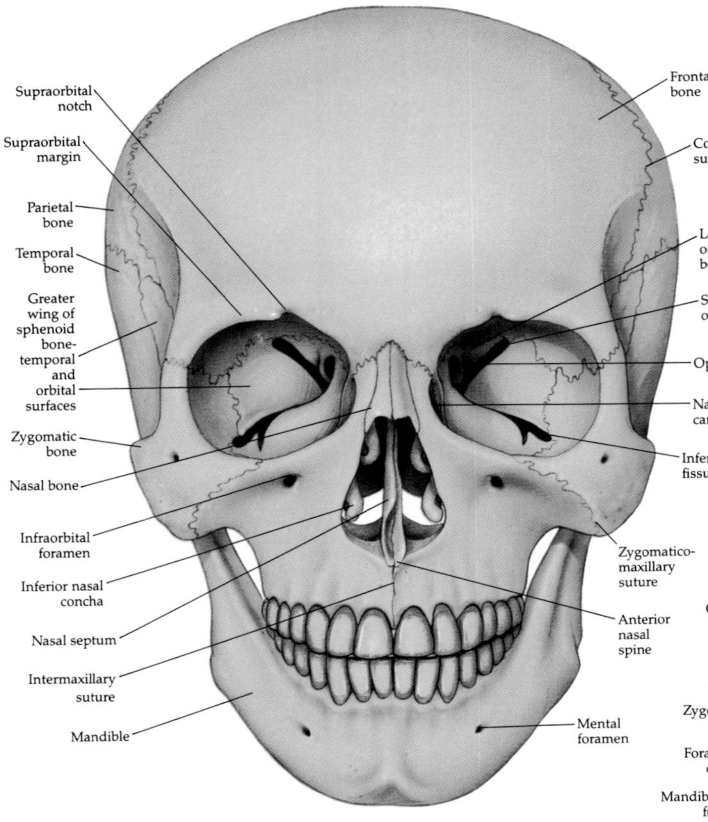

Anterior Aspect of Skull

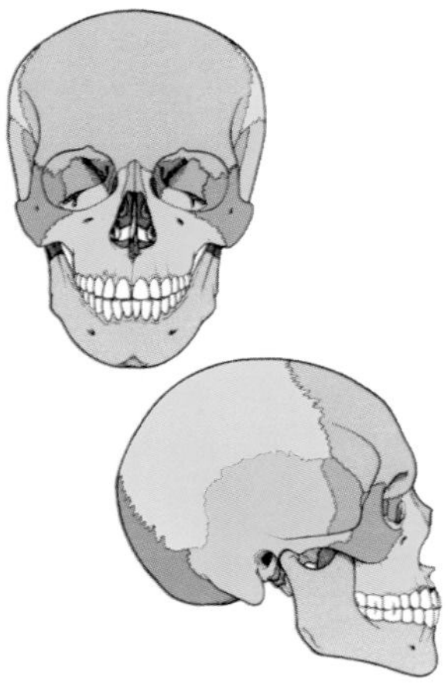

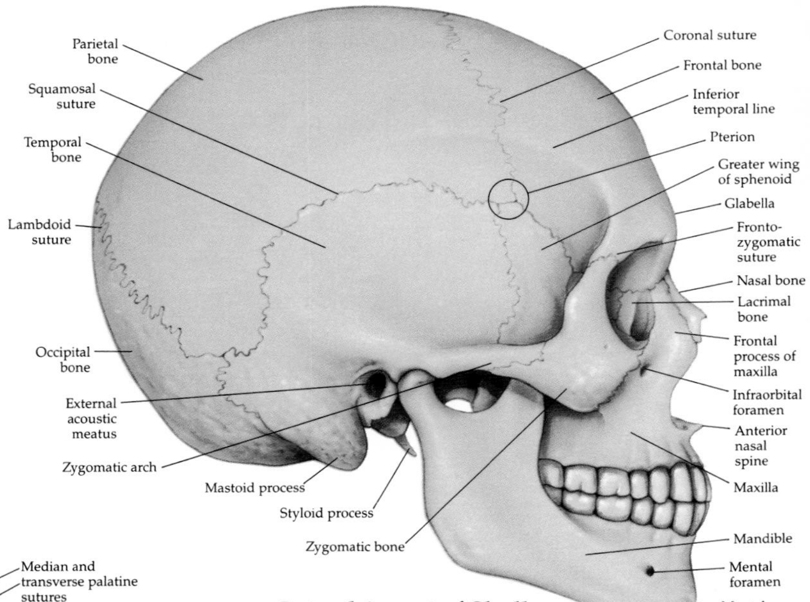

Lateral Aspect of Skull

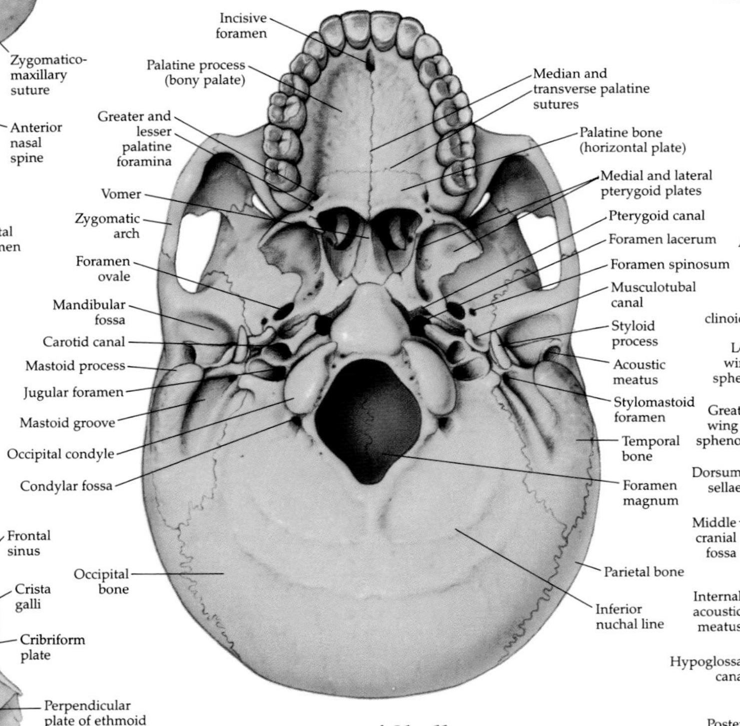

Base of Skull

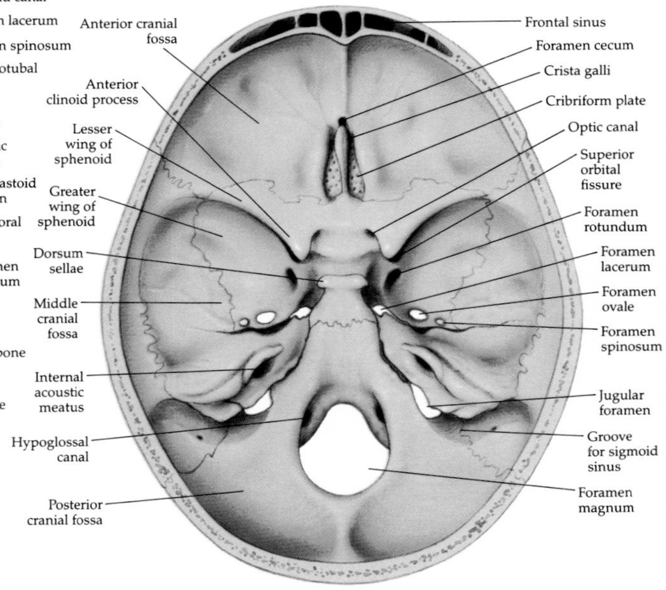

Base of Skull
(Inner Surface)

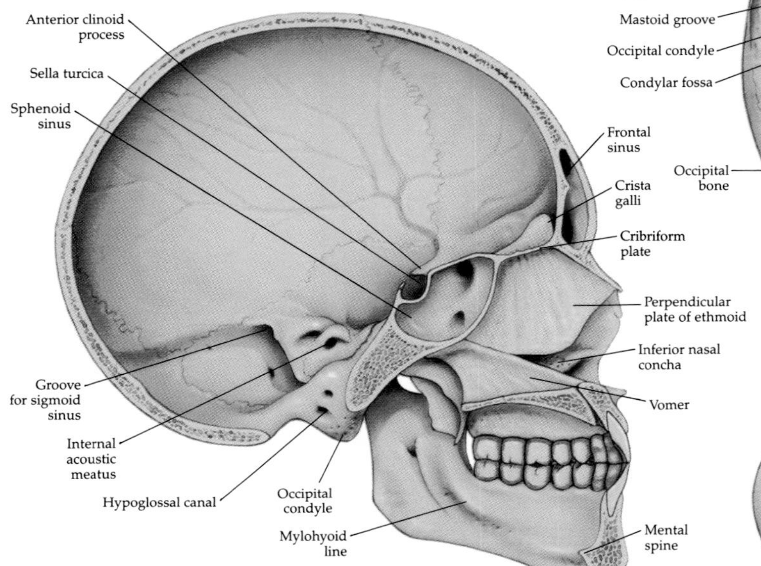

Sagittal Section Through Skull

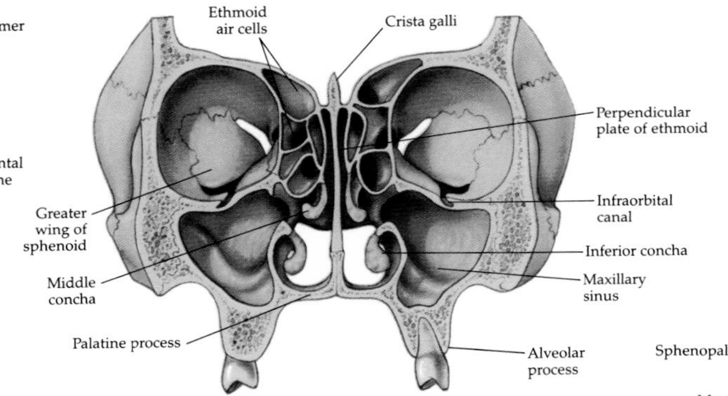

Coronal Section
Through Anterior Skull

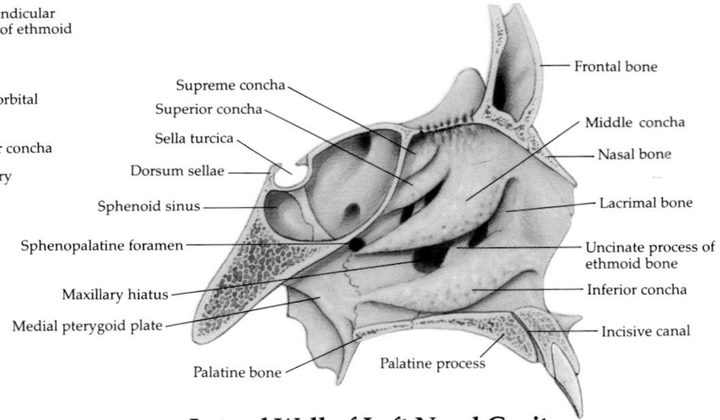

Lateral Wall of Left Nasal Cavity

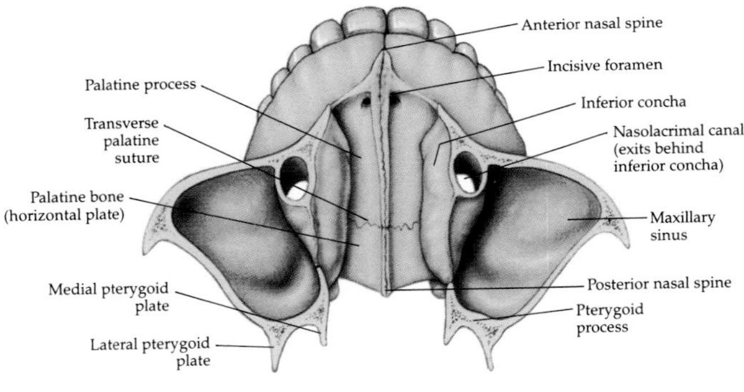

Horizontal Section Through Maxilla
(Superior View)

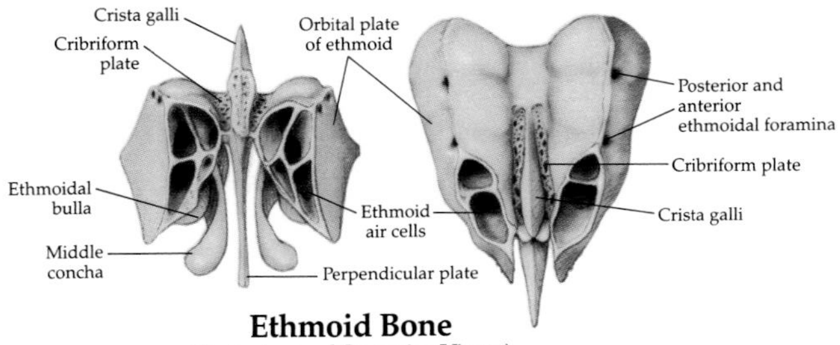

Ethmoid Bone
(Anterior and Superior Views)

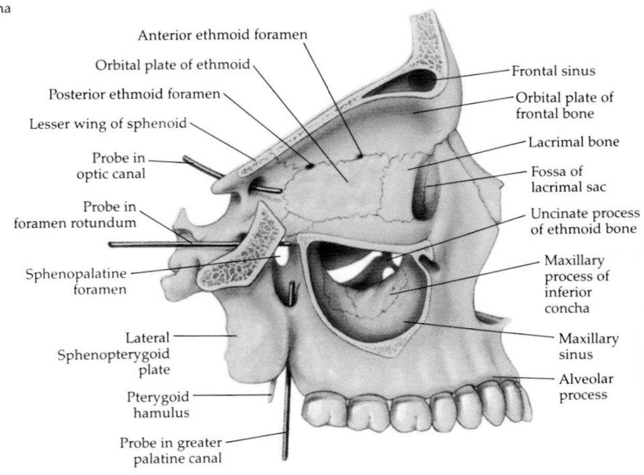

Medial Wall of Right Orbit
and Maxillary Sinus

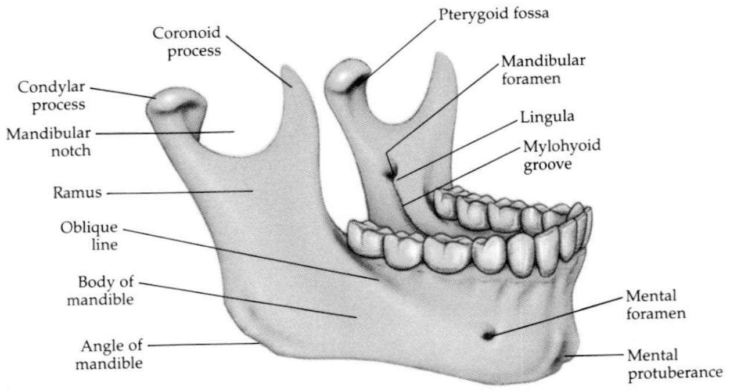

Mandible

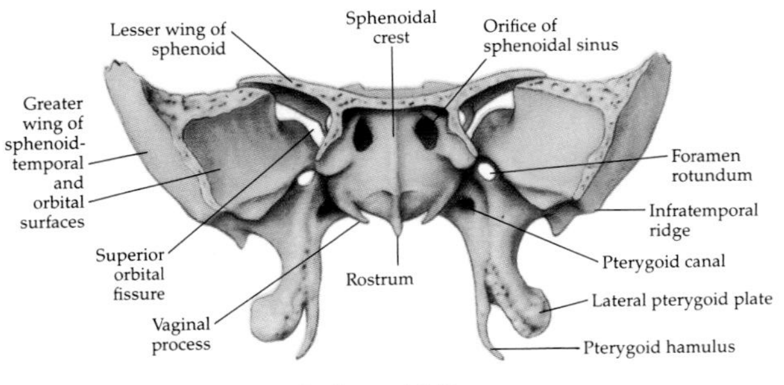

Sphenoid Bone
(Anterior View)

Medical Illustrations by William B. Westwood
in consultation with Harry Monsen, Ph.D.,
Professor of Anatomy, University of Illinois
College of Medicine at Chicago.

THE EYE

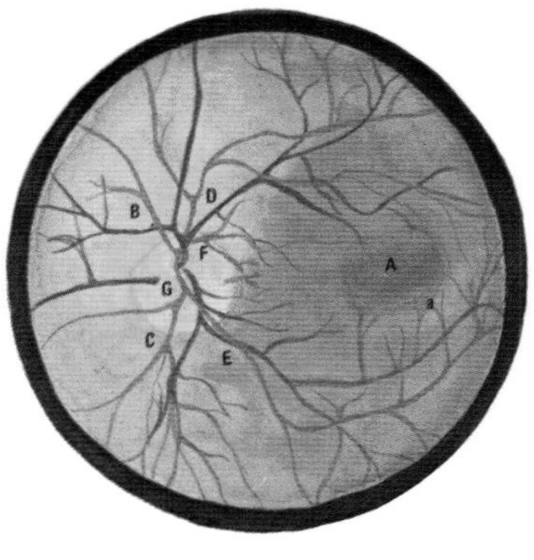

FUNDUS OF THE LEFT EYE

A. FOVEA CENTRALIS
a. MACULA LUTEA
B. SUPERIOR NASAL ARTERY
C. INFERIOR NASAL ARTERY
D. SUPERIOR TEMPORAL ARTERY
E. INFERIOR TEMPORAL ARTERY
F. CENTRAL RETINAL ARTERY
G. OPTIC NERVE

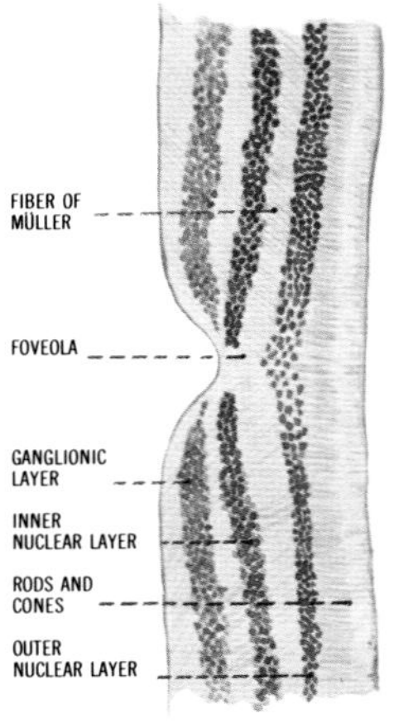

MACULA LUTEA

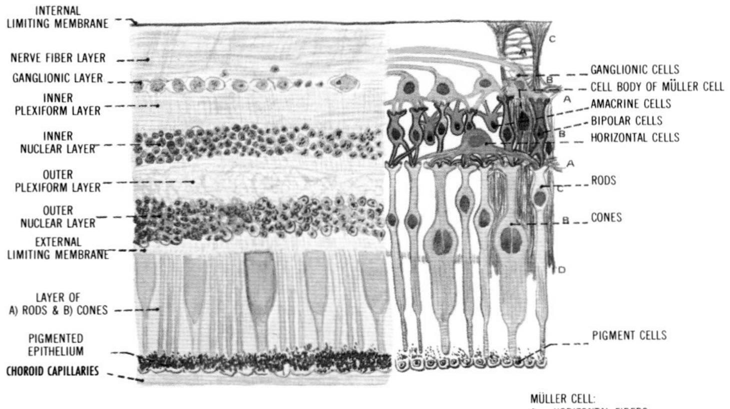

RETINA

MÜLLER CELL:
A – HORIZONTAL FIBERS
B – HONEYCOMB MESHWORK
C – RADIAL PROCESSES
D – FIBER BASKETS

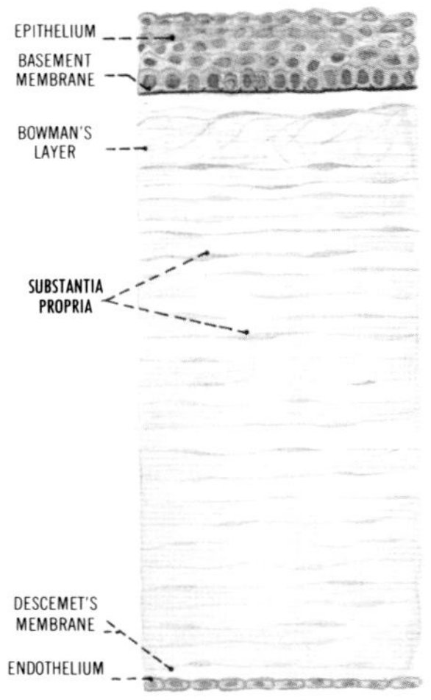

CORNEA

HORIZONTAL SECTION OF THE HUMAN EYEBALL

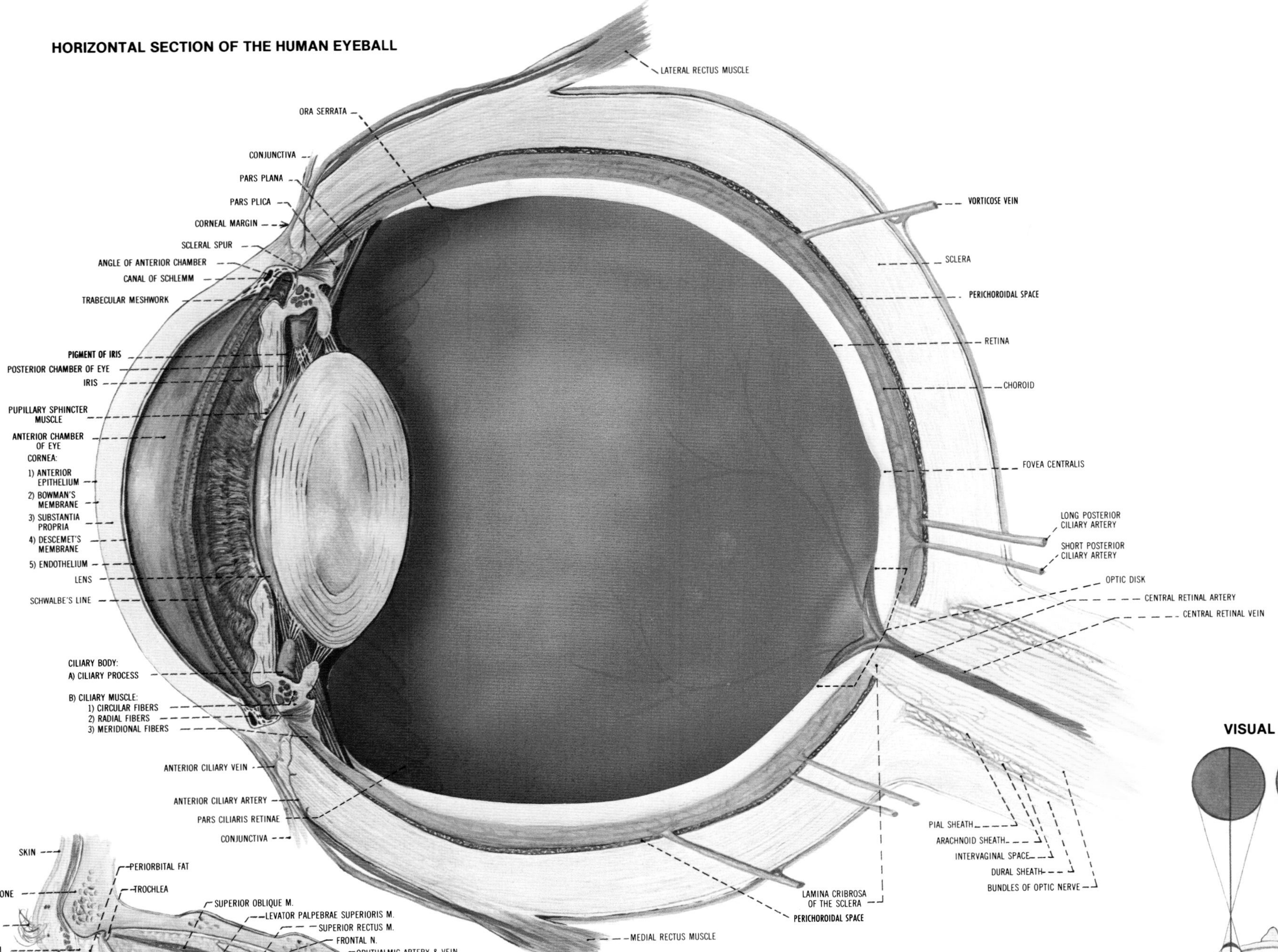

VISUAL FIELD

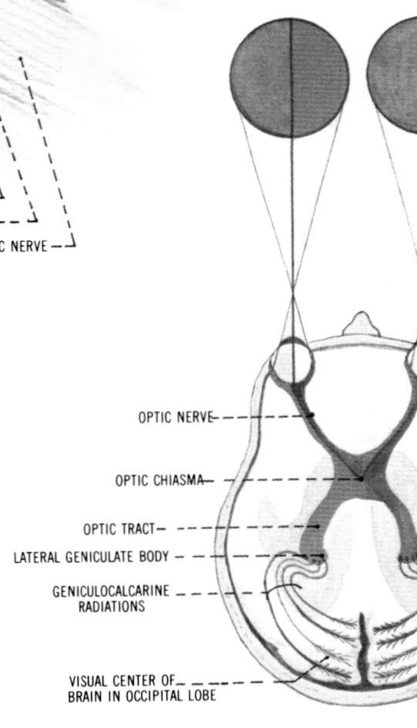

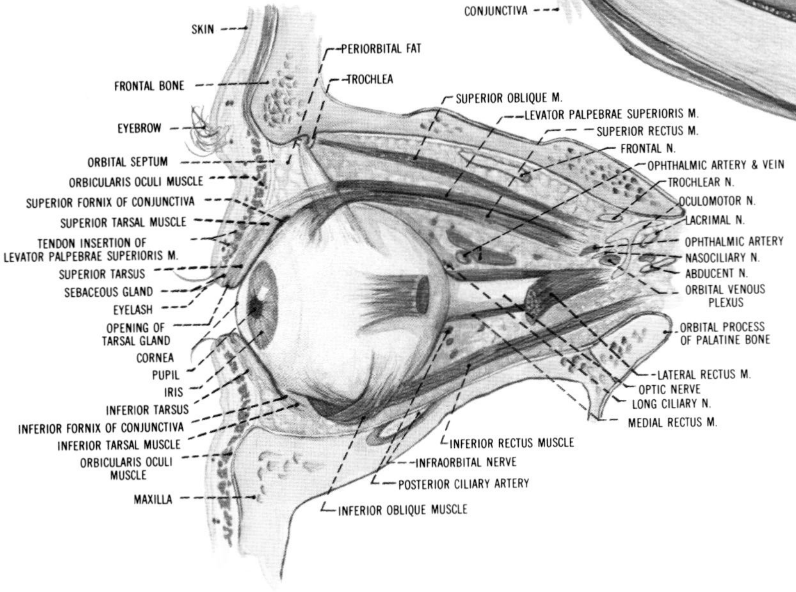

LATERAL VIEW OF LEFT EYE

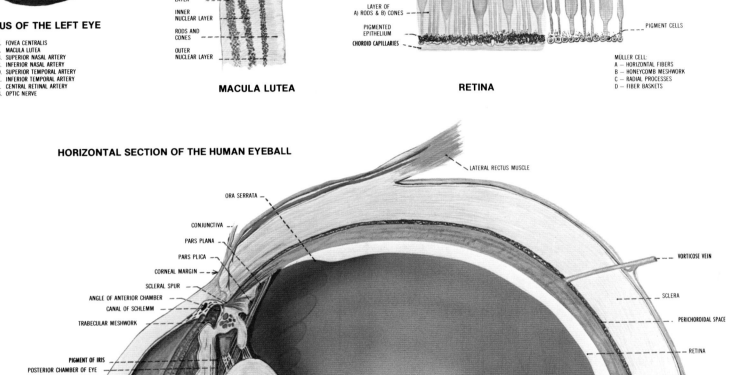

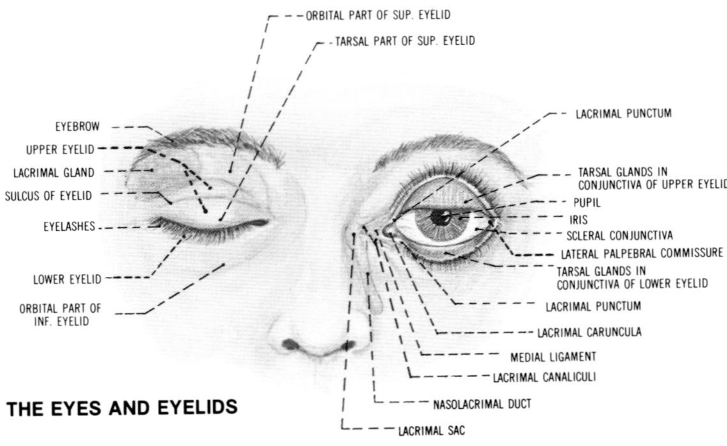

THE EYES AND EYELIDS

THE EYE: ANTERIOR AND POSTERIOR CHAMBERS

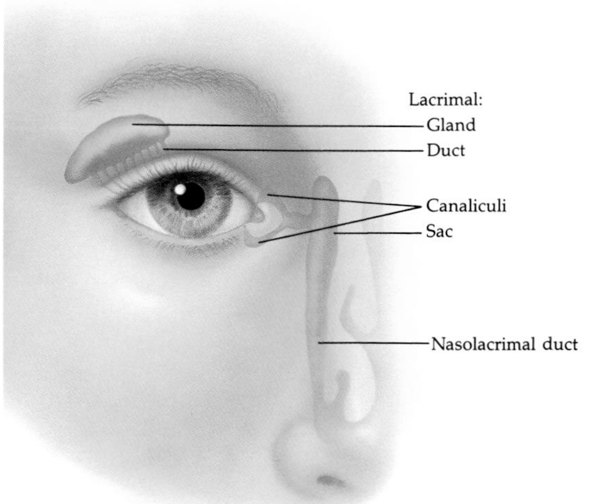

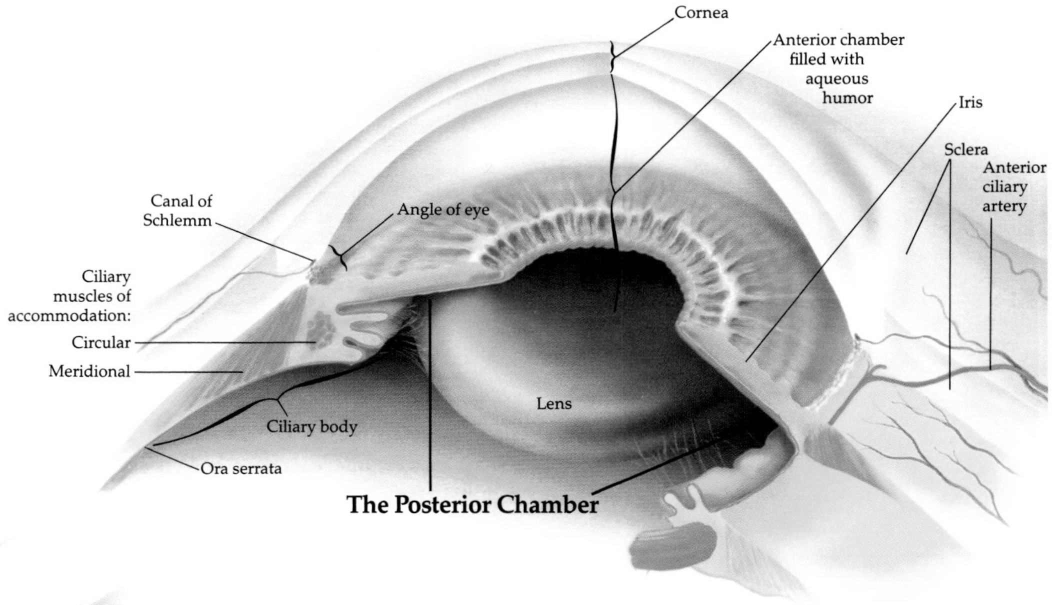

The Posterior Chamber

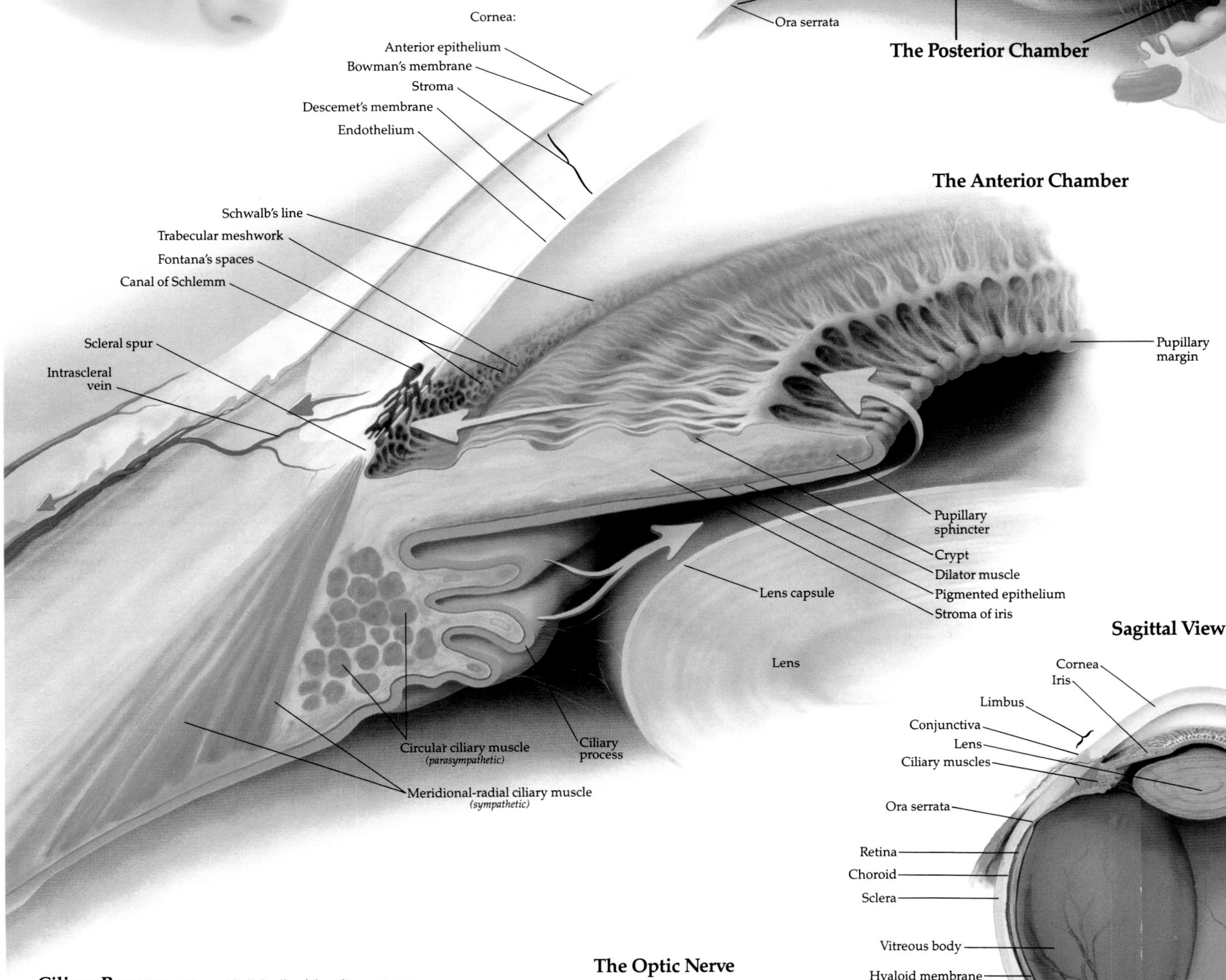

The Anterior Chamber

Sagittal View of the Eye

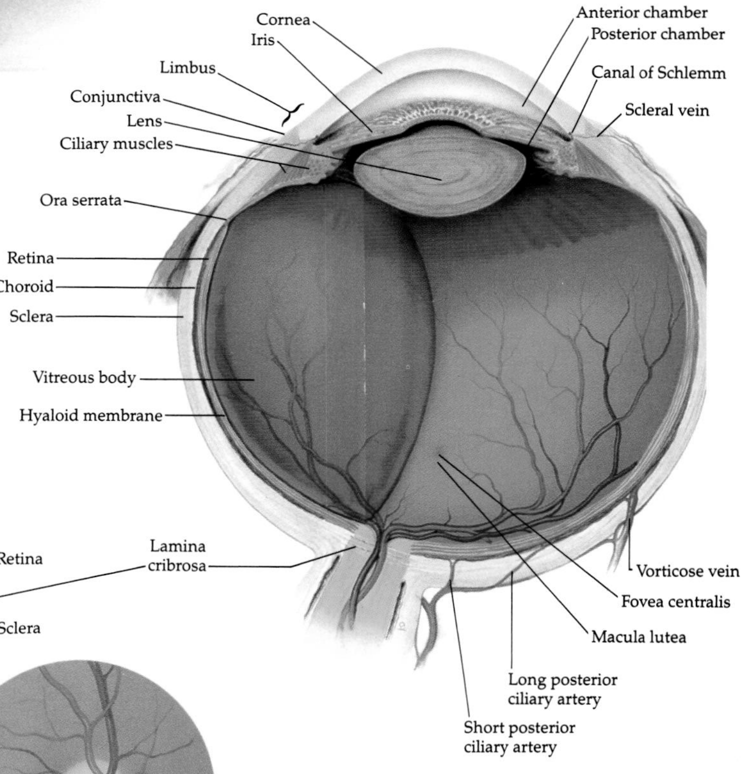

Ciliary Process

The epithelial cells of the ciliary process secrete aqueous humor into the anterior eye chamber and provide insertion sites for the zonular fibers to the lens capsule.

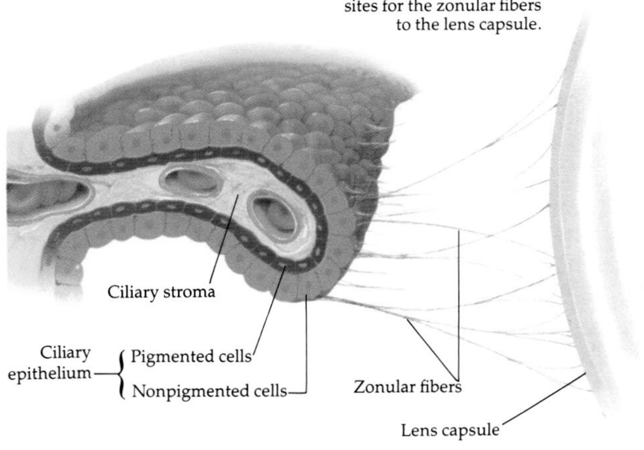

The Optic Nerve

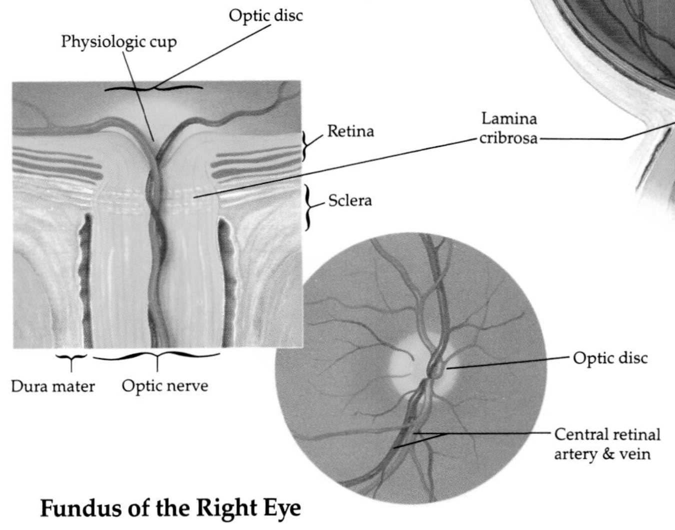

Fundus of the Right Eye

THE EAR—ORGANS OF HEARING AND BALANCE

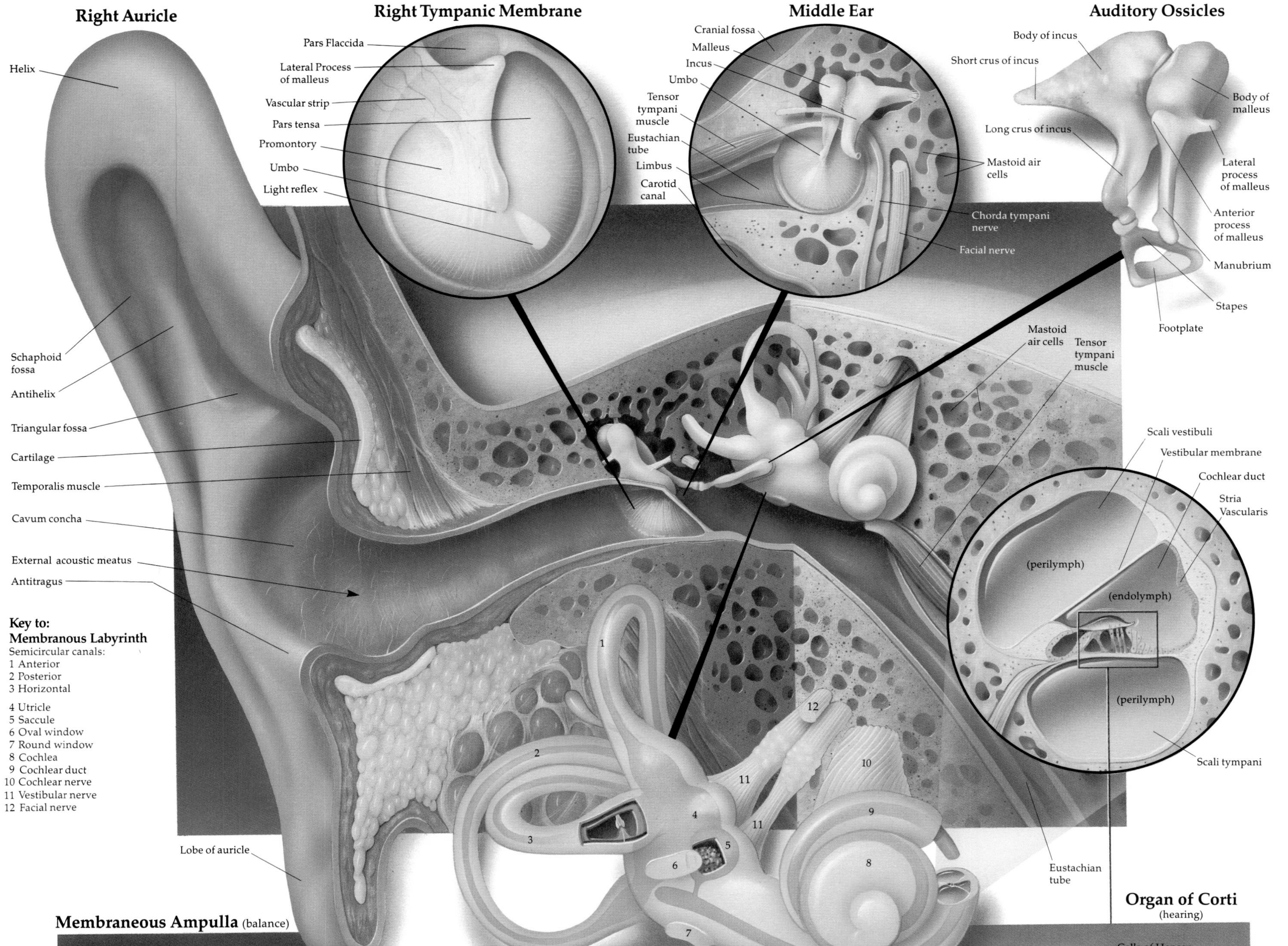

Membraneous Ampulla (balance)

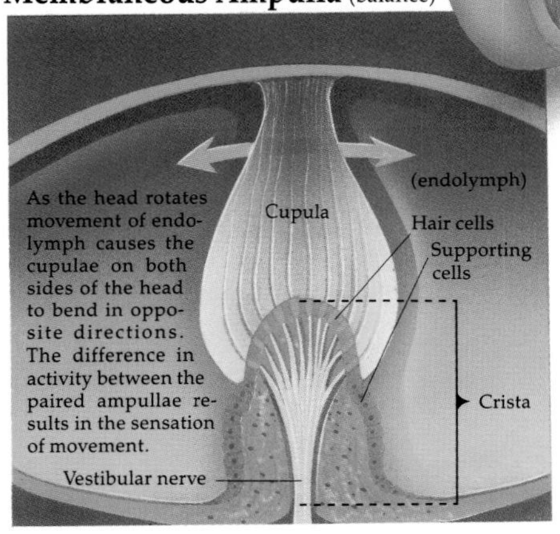

Macula of Saccule (balance)

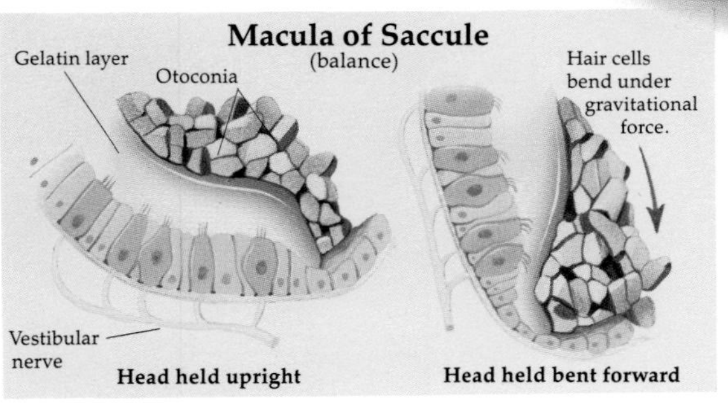

Organ of Corti (hearing)

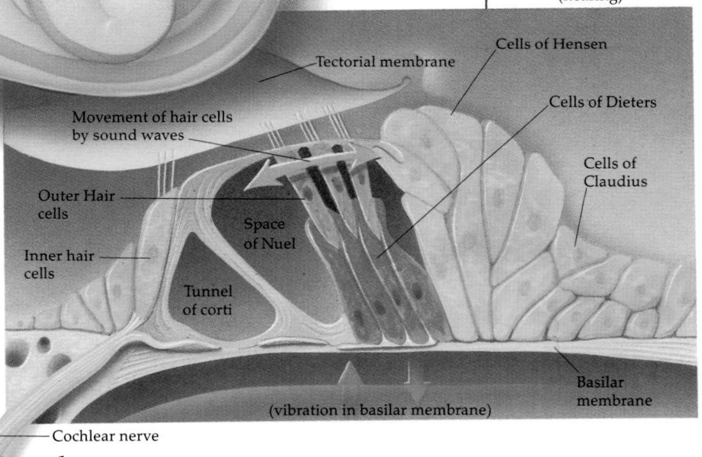

How We Hear: The Physiology of Sound

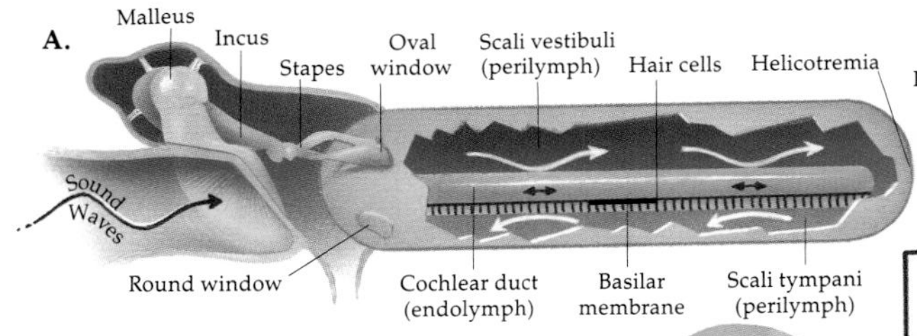

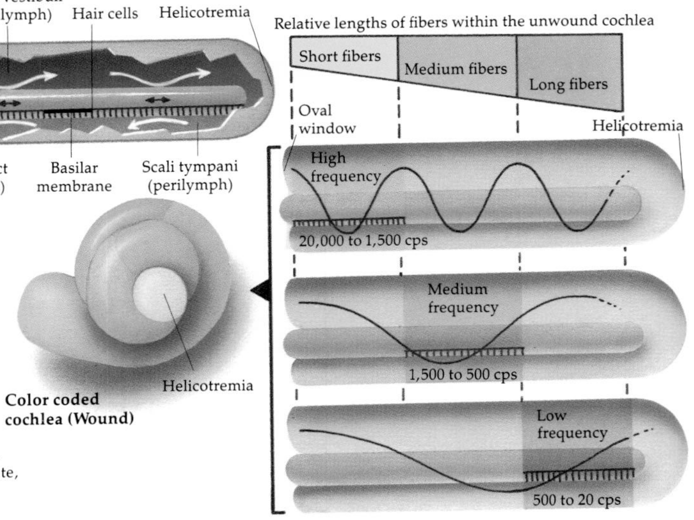

A.) Movement of the stapes in the oval window produces compression waves within the perilymph. Movement within the perilymph is transmitted to the endolymph producing vibration in the basilar membrane. These vibrations are translated by hair cells in the organ of corti into nerve impulses and sent via the cochlear nerve to the brain.

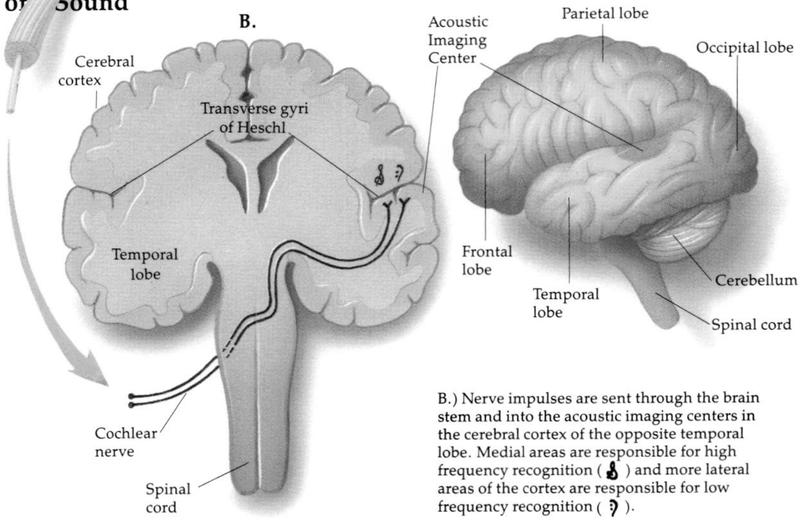

B.) Nerve impulses are sent through the brain stem and into the acoustic imaging centers in the cerebral cortex of the opposite temporal lobe. Medial areas are responsible for high frequency recognition (♪) and more lateral areas of the cortex are responsible for low frequency recognition (?).

Medical illustrations by Keith Kasnot in consultation with C. Phillip Daspit M.D. F.A.C.S. Head of Neurotology, Barrow Neurological Institute, Phoenix, Arizona

EAR, NOSE AND THROAT

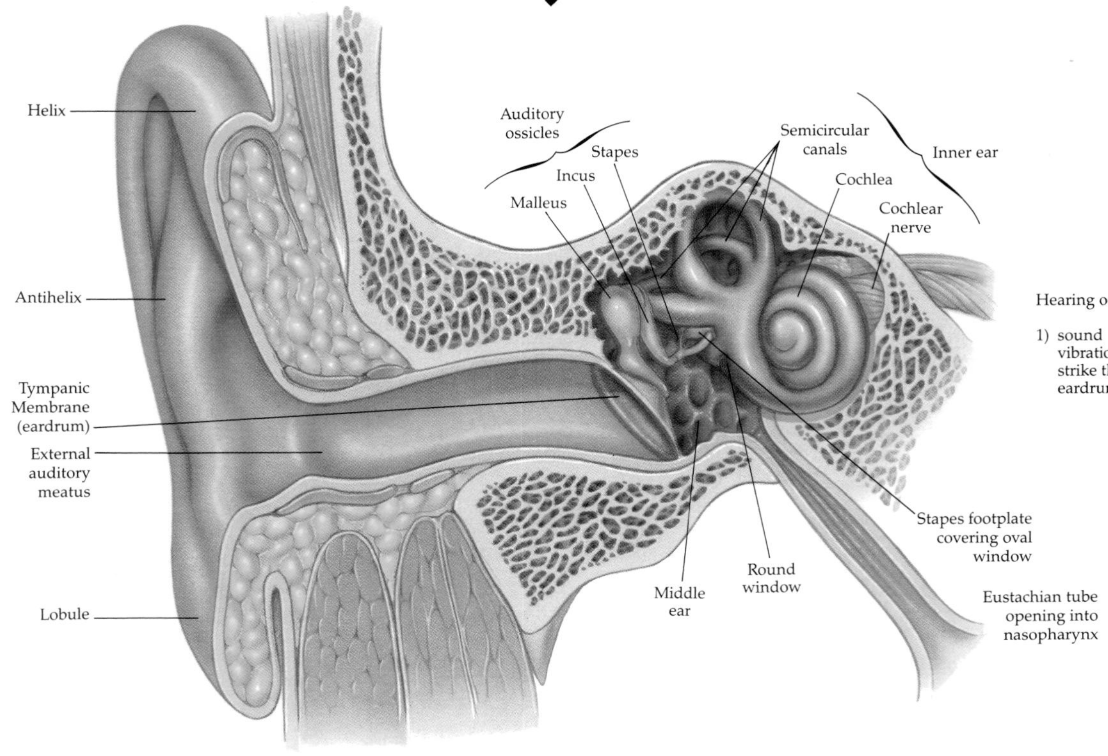

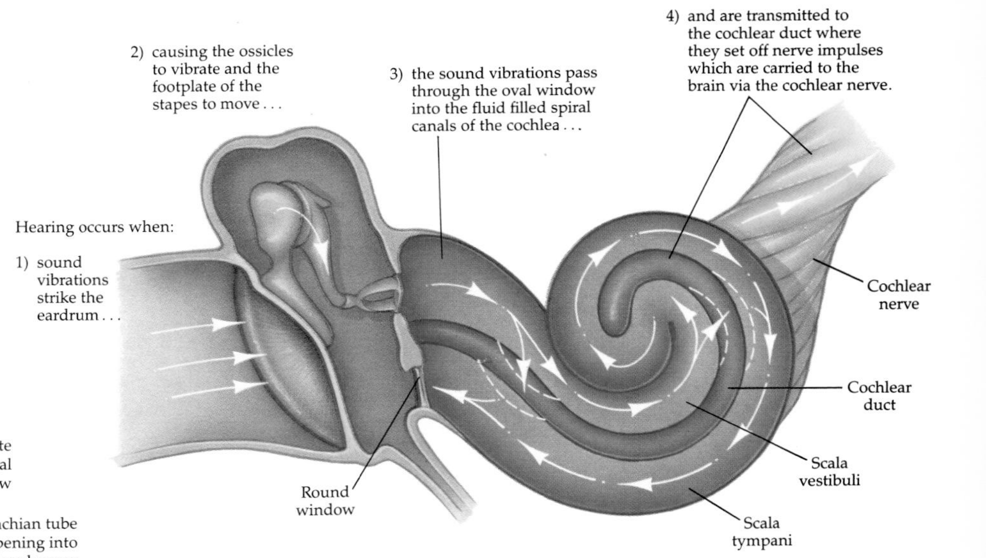

Middle & Inner Ear

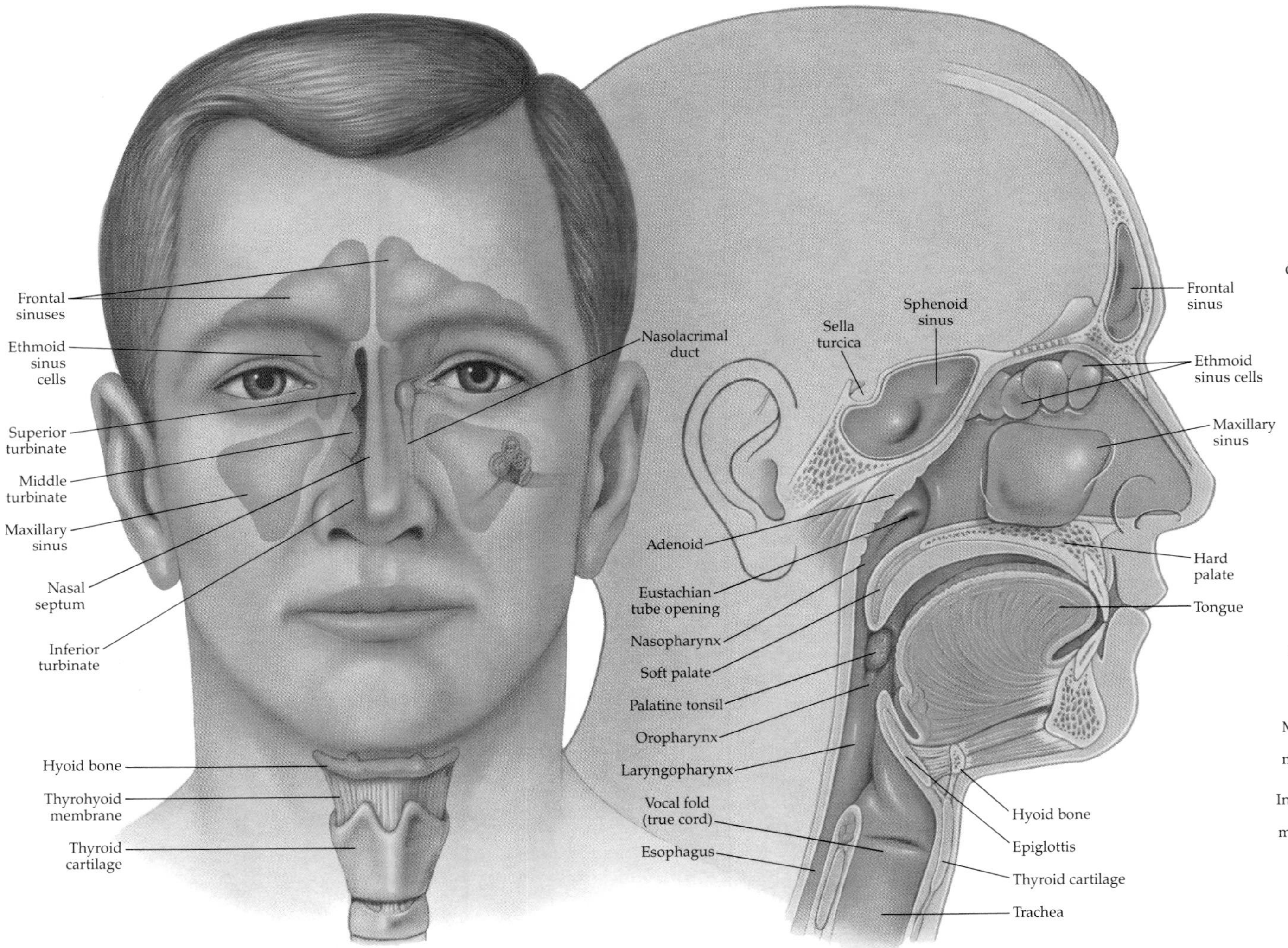

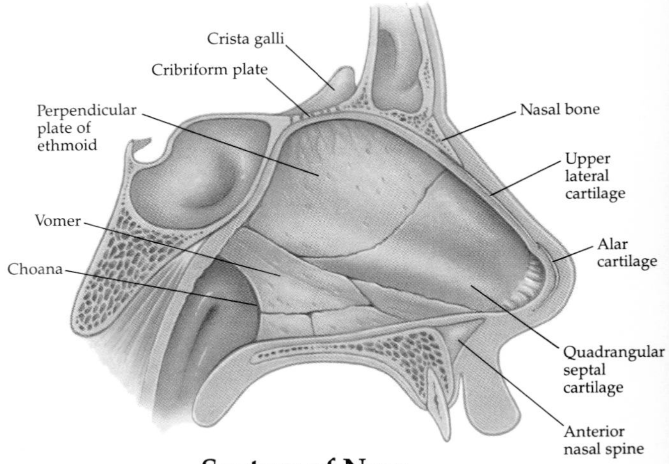

Septum of Nose

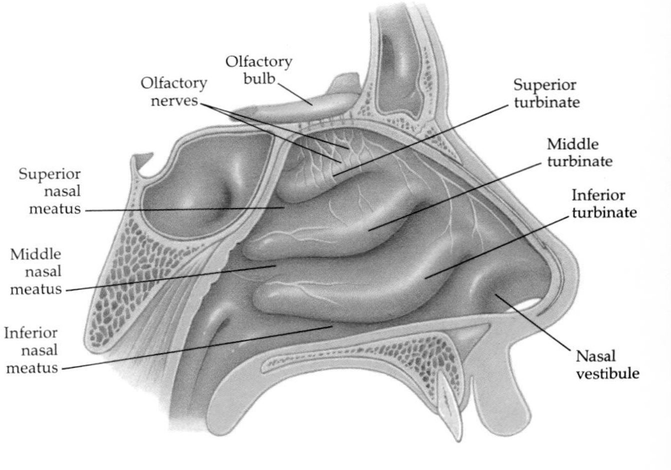

Lateral Wall of Nose

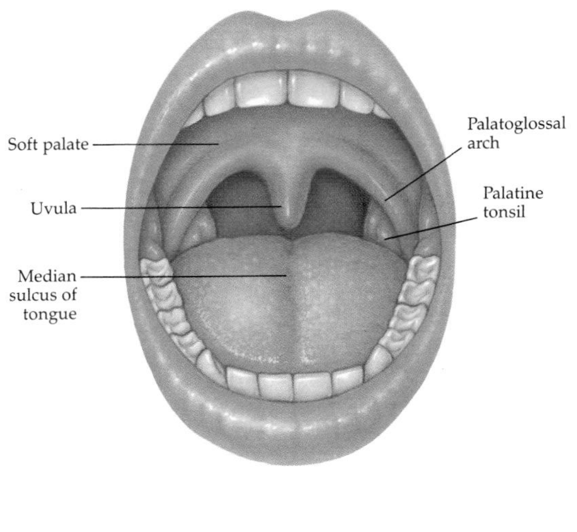

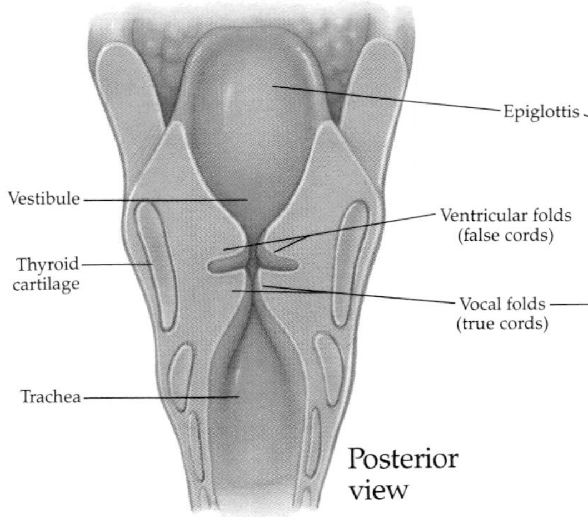

Posterior view

Normal Larynx

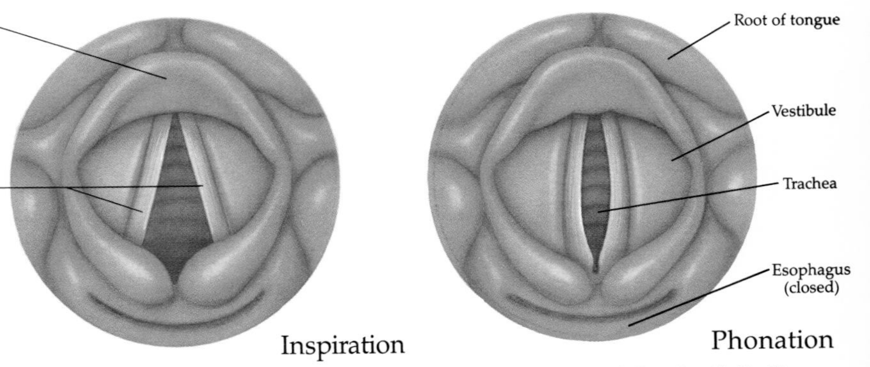

Inspiration

Phonation

Medical Illustrations by William B. Westwood in consultation with Eugene B. Kern, M.D., Professor of Otorhinolaryngology.

THE SKIN

NAIL

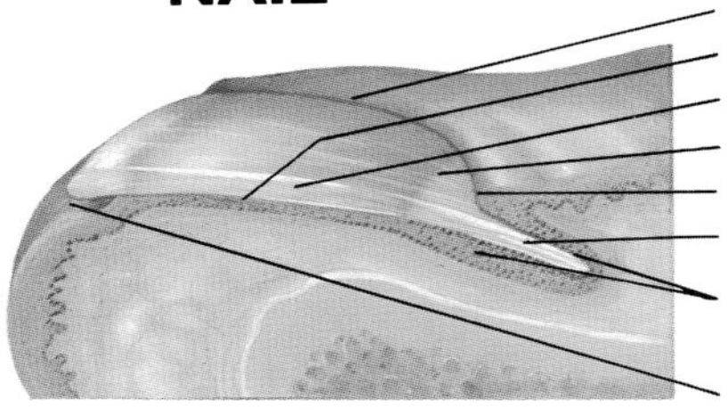
Lateral nail fold
Nail bed
Nail plate
Lunula
Eponychium
Nail root
Nail Matrix
(Germinal layer)
Hyponychium

HAIR

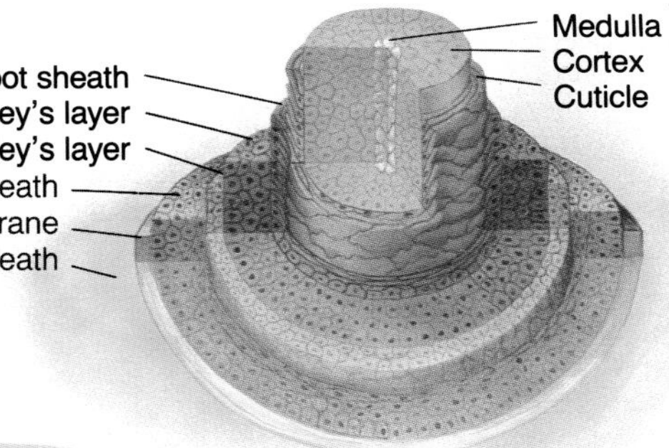
Medulla
Cortex
Cuticle
Hair Shaft
Inner root sheath
Huxley's layer
Henley's layer
Outer root sheath
Glassy Membrane
Connective tissue sheath

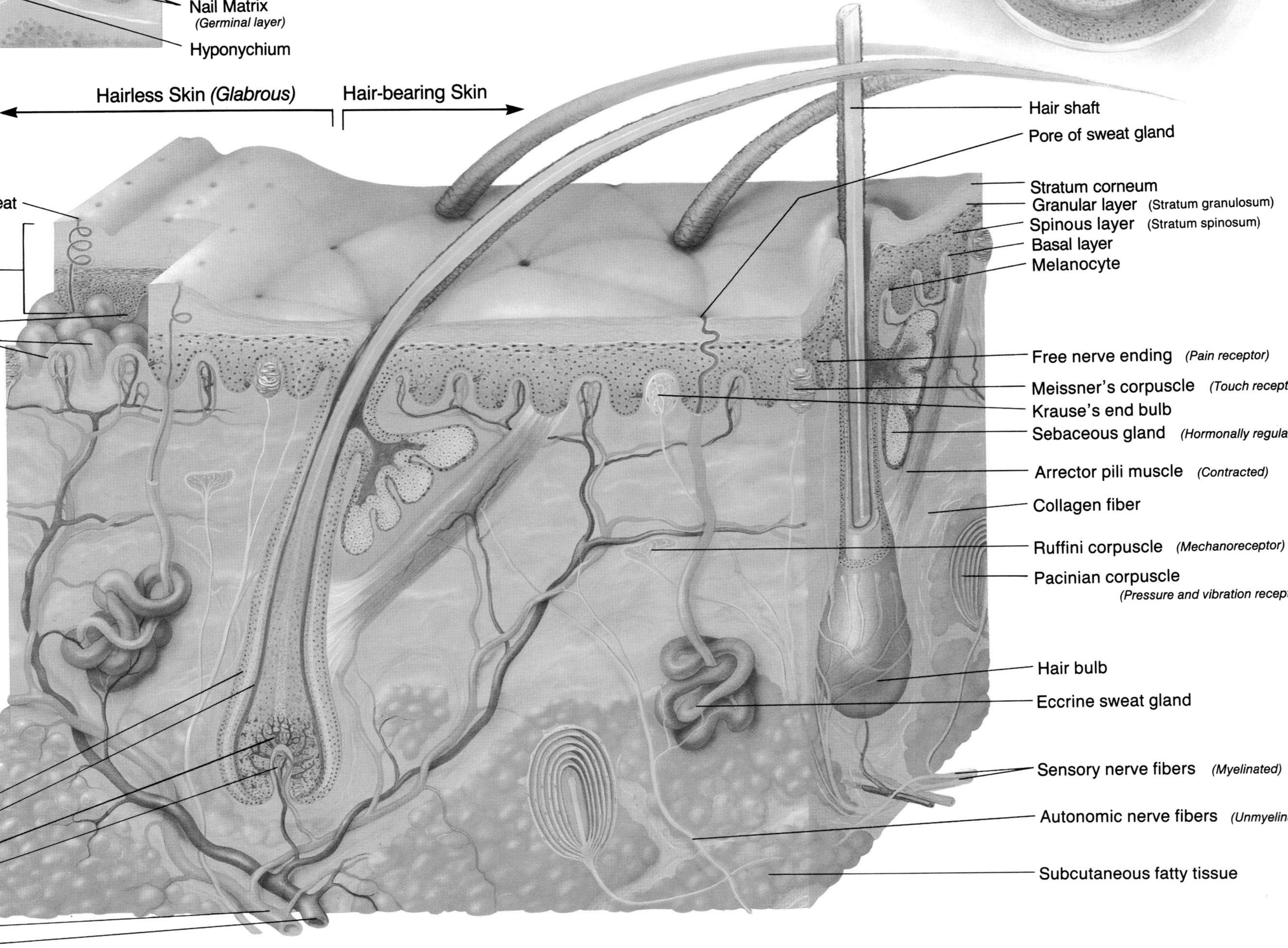
Hairless Skin (Glabrous)
Hair-bearing Skin
Sweat
Epidermis
Epidermal rete ridge
Dermal papillae
Dermis
Subcutaneous Tissue
Outer root sheath
Inner root sheath
Matrix cells
Hair papilla
Artery
Vein
Hair shaft
Pore of sweat gland
Stratum corneum
Granular layer (Stratum granulosum)
Spinous layer (Stratum spinosum)
Basal layer
Melanocyte
Free nerve ending (Pain receptor)
Meissner's corpuscle (Touch receptor)
Krause's end bulb
Sebaceous gland (Hormonally regulated)
Arrector pili muscle (Contracted)
Collagen fiber
Ruffini corpuscle (Mechanoreceptor)
Pacinian corpuscle
(Pressure and vibration receptor)
Hair bulb
Eccrine sweat gland
Sensory nerve fibers (Myelinated)
Autonomic nerve fibers (Unmyelinated)
Subcutaneous fatty tissue

TYPES OF SKIN LESIONS

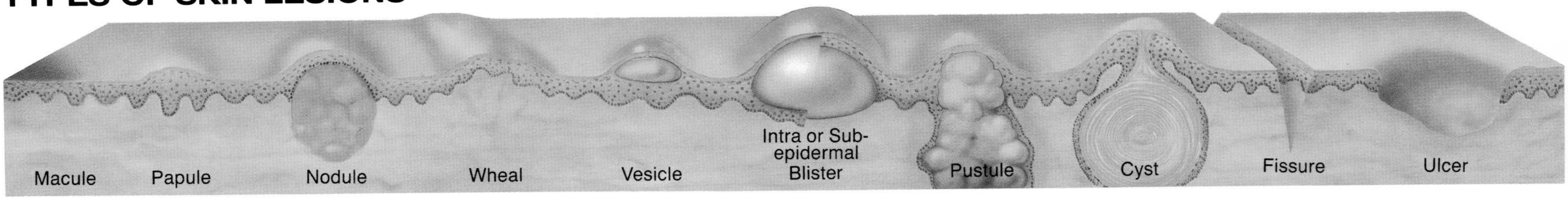
Macule
Papule
Nodule
Wheal
Vesicle
Intra or Sub-epidermal Blister
Pustule
Cyst
Fissure
Ulcer

COMMON SKIN AILMENTS

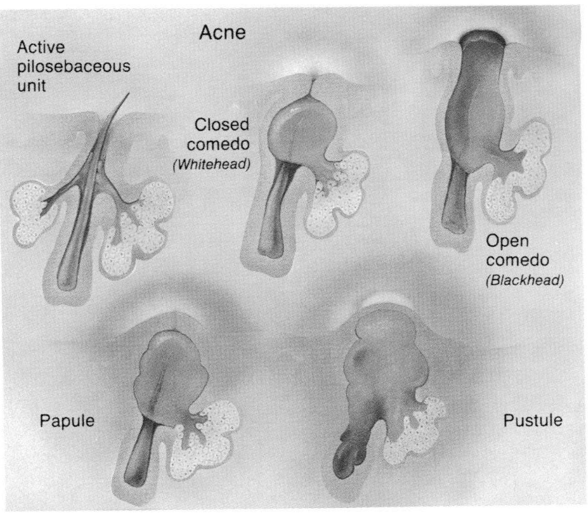
Acne
Active pilosebaceous unit
Closed comedo
(Whitehead)
Open comedo
(Blackhead)
Papule
Pustule

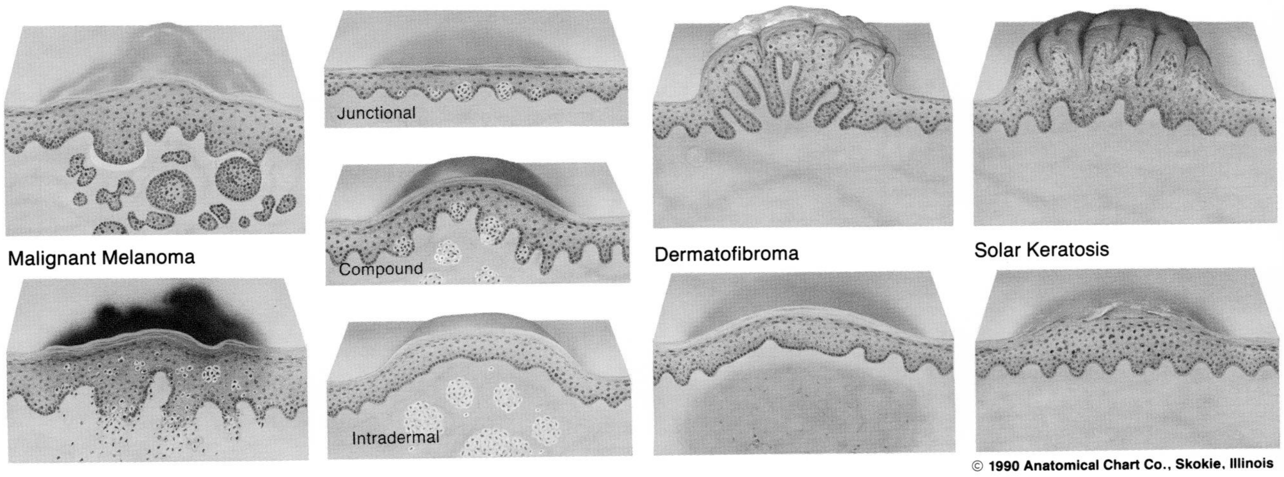
Basal Cell Carcinoma
Malignant Melanoma
Melanocytic Nevus (Mole)
Junctional
Compound
Intradermal
Papilloma Virus Infection (Wart)
Dermatofibroma
Seborrhoeic Keratosis
Solar Keratosis

THE HUMAN HAIR

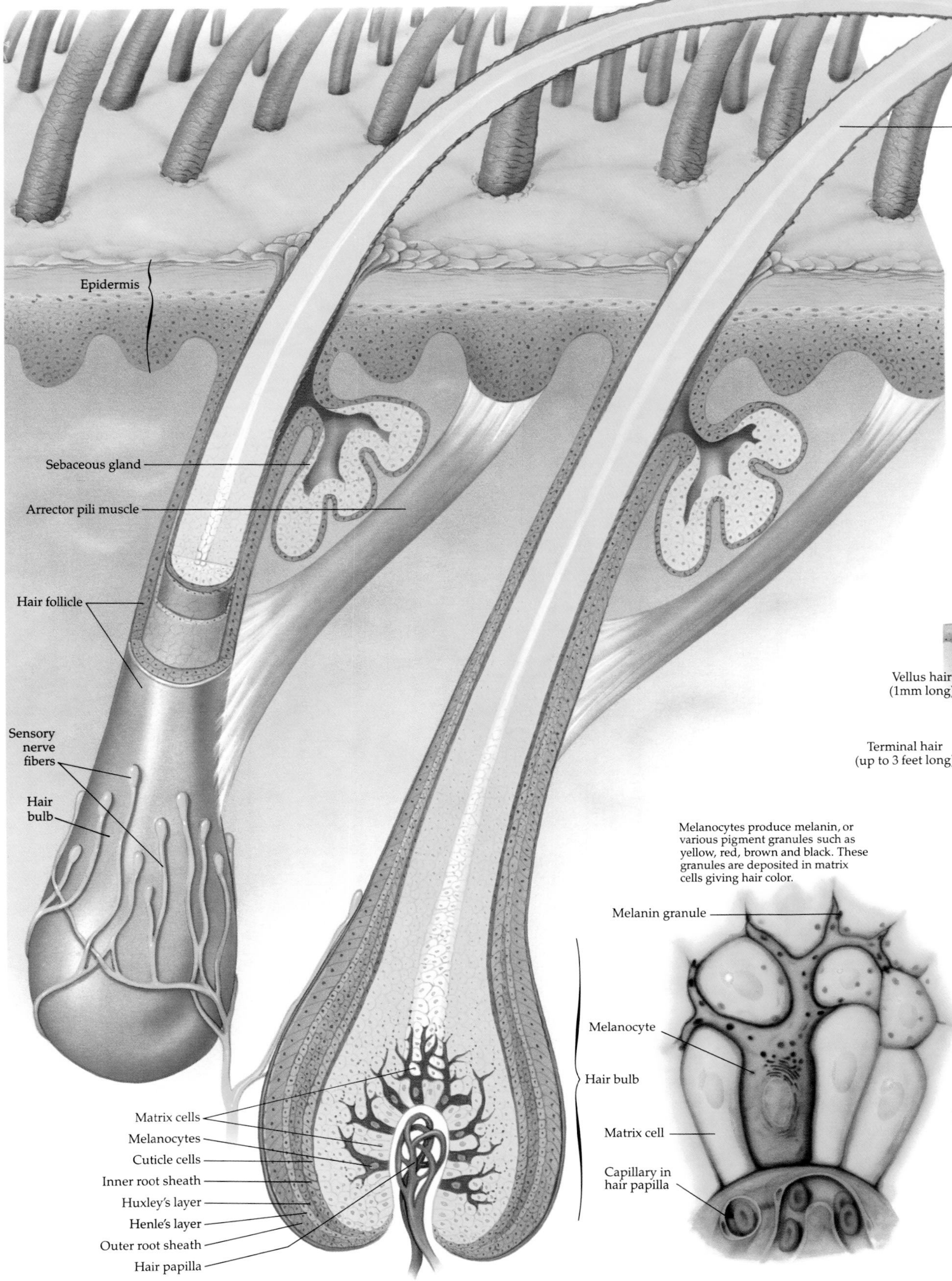

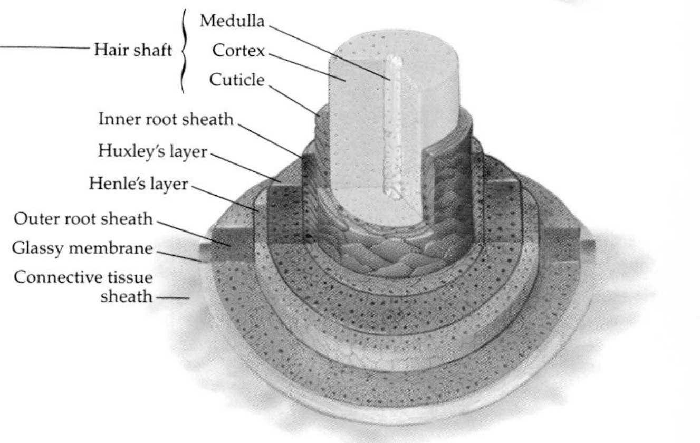

Types of Scalp Hair

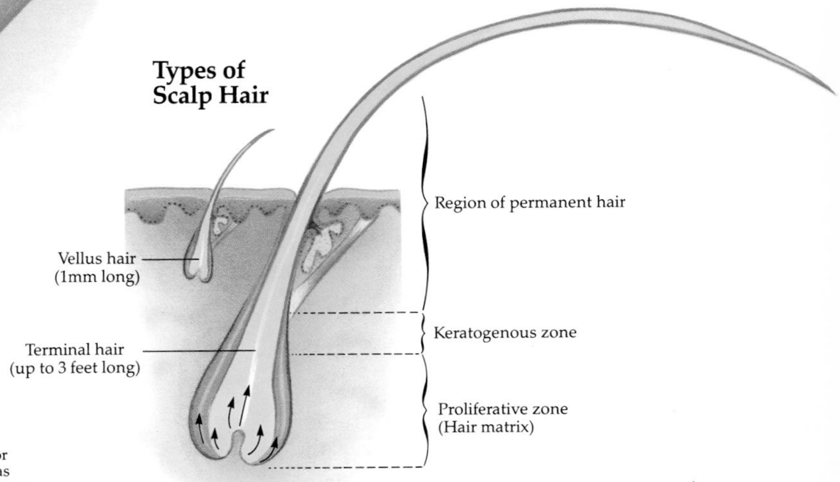

Melanocytes produce melanin, or various pigment granules such as yellow, red, brown and black. These granules are deposited in matrix cells giving hair color.

Hair Fiber Characteristics

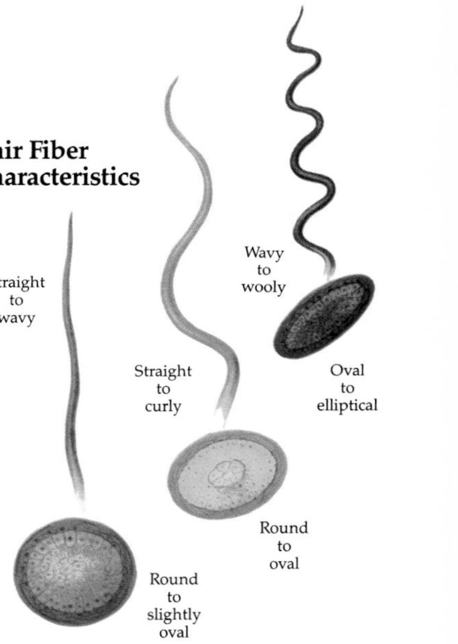

Hair Analysis

Trace elements are incorporated into hair protein structure from the blood stream as the cells grow from the hair bulb outward. Analysis of hair fibers may reveal dietary deficiency, systemic intoxication, environmental exposure to heavy-metals and the presence of therapeutic or abused drugs.

Miniaturization of Hair Follicles in Baldness

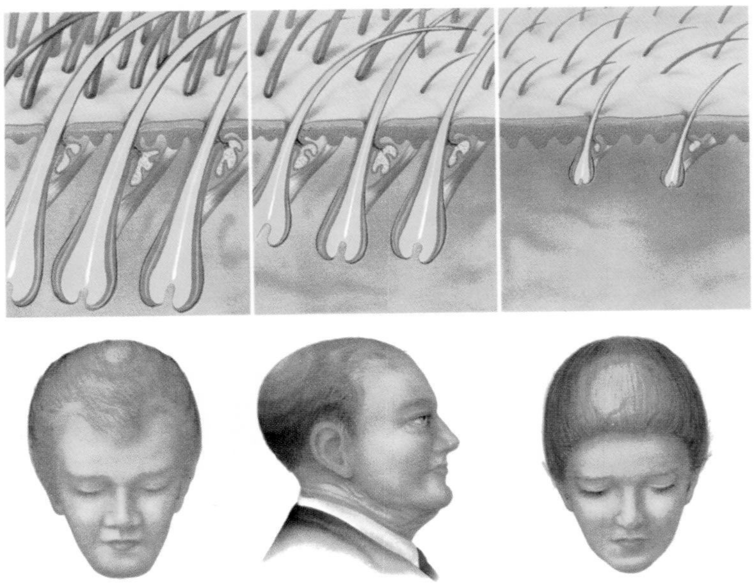

Hair-Growth Cycle

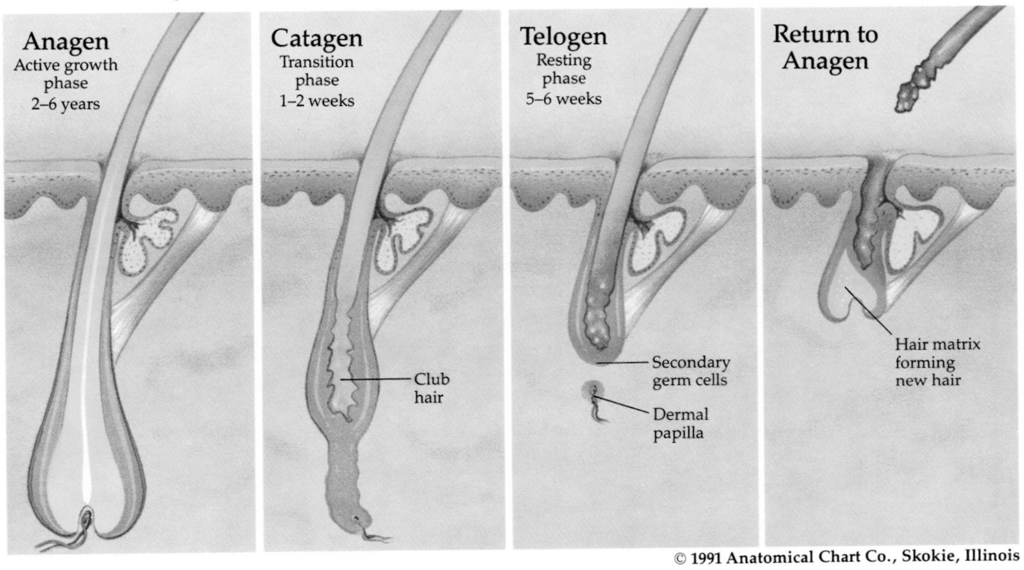

SHOULDER AND ELBOW

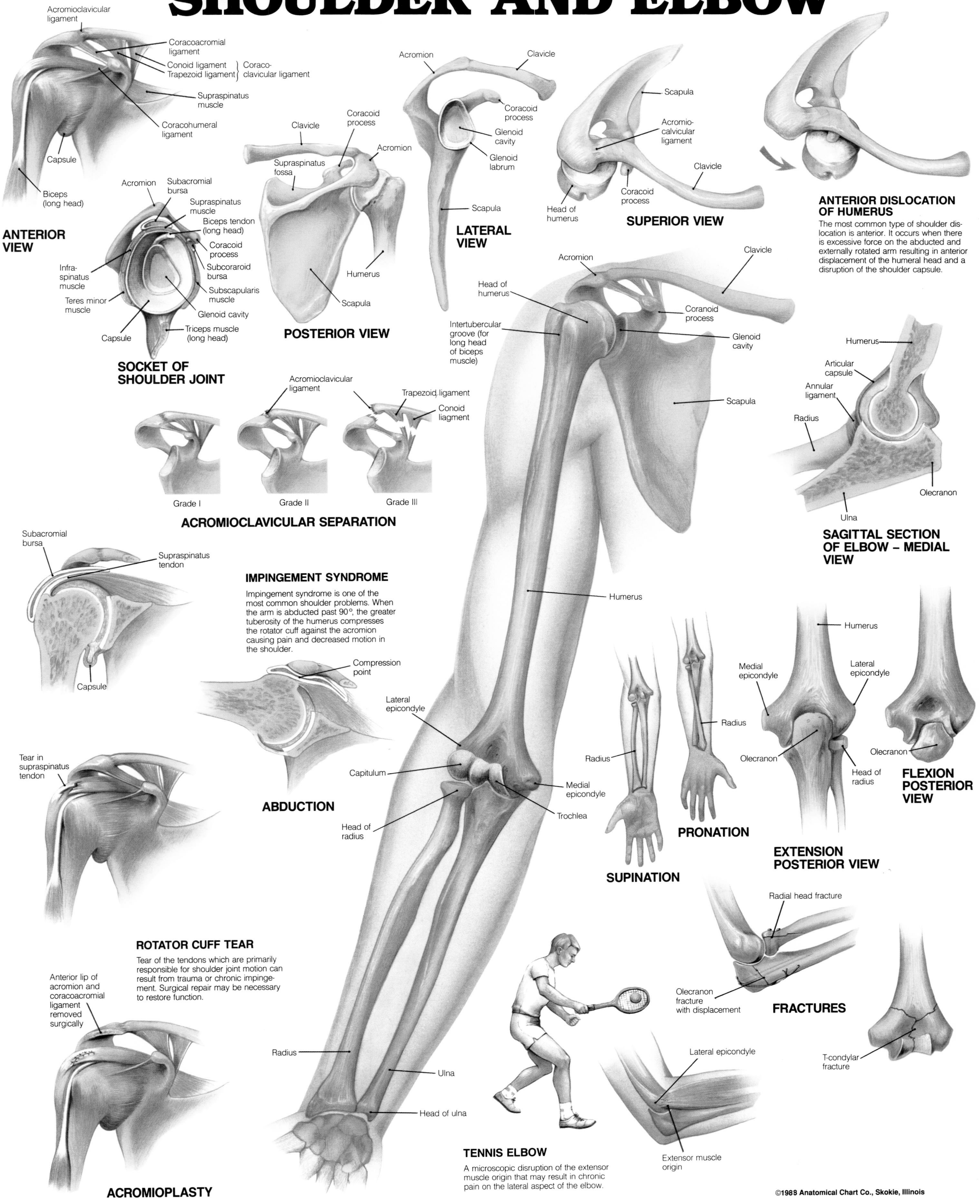
Acromioclavicular ligament
Coracoacromial ligament
Conoid ligament
Trapezoid ligament
Coraco-clavicular ligament
Supraspinatus muscle
Coracohumeral ligament
Capsule
Biceps (long head)
ANTERIOR VIEW
Acromion
Subacromial bursa
Supraspinatus muscle
Biceps tendon (long head)
Coracoid process
Subcoraroid bursa
Subscapularis muscle
Glenoid cavity
Triceps muscle (long head)
Infra-spinatus muscle
Teres minor muscle
Capsule
SOCKET OF SHOULDER JOINT
Clavicle
Coracoid process
Acromion
Supraspinatus fossa
Humerus
Scapula
POSTERIOR VIEW
Acromion
Clavicle
Coracoid process
Glenoid cavity
Glenoid labrum
Scapula
LATERAL VIEW
Scapula
Acromio-calvicular ligament
Clavicle
Coracoid process
Head of humerus
SUPERIOR VIEW
ANTERIOR DISLOCATION OF HUMERUS
The most common type of shoulder dis-location is anterior. It occurs when there is excessive force on the abducted and externally rotated arm resulting in anterior displacement of the humeral head and a disruption of the shoulder capsule.
Acromioclavicular ligament
Trapezoid ligament
Conoid liagment
Grade I
Grade II
Grade III
ACROMIOCLAVICULAR SEPARATION
Acromion
Clavicle
Head of humerus
Coranoid process
Glenoid cavity
Intertubercular groove (for long head of biceps muscle)
Scapula
Humerus
Lateral epicondyle
Capitulum
Medial epicondyle
Trochlea
Head of radius
Radius
Ulna
Head of ulna
Humerus
Articular capsule
Annular ligament
Radius
Olecranon
Ulna
SAGITTAL SECTION OF ELBOW – MEDIAL VIEW
Subacromial bursa
Supraspinatus tendon
Capsule
IMPINGEMENT SYNDROME
Impingement syndrome is one of the most common shoulder problems. When the arm is abducted past 90°, the greater tuberosity of the humerus compresses the rotator cuff against the acromion causing pain and decreased motion in the shoulder.
Compression point
ABDUCTION
Tear in supraspinatus tendon
ROTATOR CUFF TEAR
Tear of the tendons which are primarily responsible for shoulder joint motion can result from trauma or chronic impinge-ment. Surgical repair may be necessary to restore function.
Anterior lip of acromion and coracoacromial ligament removed surgically
ACROMIOPLASTY
Radius
SUPINATION
Radius
PRONATION
Humerus
Medial epicondyle
Lateral epicondyle
Olecranon
Head of radius
EXTENSION POSTERIOR VIEW
Olecranon
FLEXION POSTERIOR VIEW
Radial head fracture
Olecranon fracture with displacement
FRACTURES
T-condylar fracture
Lateral epicondyle
Extensor muscle origin
TENNIS ELBOW
A microscopic disruption of the extensor muscle origin that may result in chronic pain on the lateral aspect of the elbow.
©1988 Anatomical Chart Co., Skokie, Illinois

HAND AND WRIST

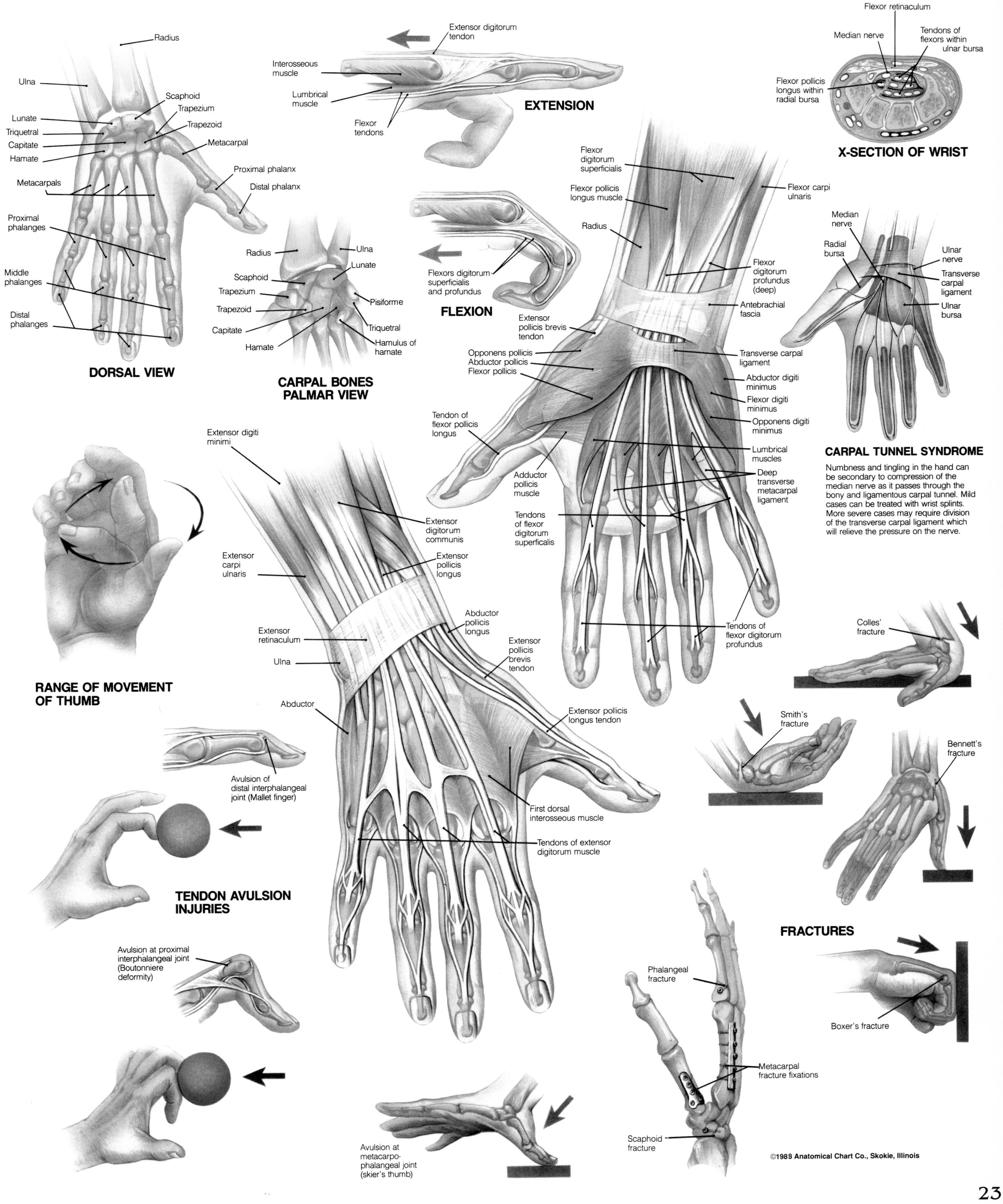
Radius
Ulna
Scaphoid
Trapezium
Lunate
Triquetral
Trapezoid
Capitate
Metacarpal
Hamate
Proximal phalanx
Metacarpals
Distal phalanx
Proximal phalanges
Middle phalanges
Distal phalanges
DORSAL VIEW
Radius
Ulna
Lunate
Scaphoid
Trapezium
Pisiforme
Trapezoid
Triquetral
Capitate
Hamulus of hamate
Hamate
CARPAL BONES PALMAR VIEW
Extensor digitorum tendon
Interosseous muscle
Lumbrical muscle
Flexor tendons
EXTENSION
Flexors digitorum superficialis and profundus
FLEXION
Flexor retinaculum
Median nerve
Tendons of flexors within ulnar bursa
Flexor pollicis longus within radial bursa
X-SECTION OF WRIST
Flexor digitorum superficialis
Flexor pollicis longus muscle
Flexor carpi ulnaris
Radius
Flexor digitorum profundus (deep)
Antebrachial fascia
Extensor pollicis brevis tendon
Opponens pollicis
Abductor pollicis
Flexor pollicis
Transverse carpal ligament
Abductor digiti minimus
Flexor digiti minimus
Opponens digiti minimus
Tendon of flexor pollicis longus
Lumbrical muscles
Adductor pollicis muscle
Deep transverse metacarpal ligament
Tendons of flexor digitorum superficalis
Tendons of flexor digitorum profundus
Median nerve
Radial bursa
Ulnar nerve
Transverse carpal ligament
Ulnar bursa
CARPAL TUNNEL SYNDROME
Numbness and tingling in the hand can be secondary to compression of the median nerve as it passes through the bony and ligamentous carpal tunnel. Mild cases can be treated with wrist splints. More severe cases may require division of the transverse carpal ligament which will relieve the pressure on the nerve.
RANGE OF MOVEMENT OF THUMB
Extensor digiti minimi
Extensor digitorum communis
Extensor carpi ulnaris
Extensor pollicis longus
Abductor pollicis longus
Extensor retinaculum
Extensor pollicis brevis tendon
Ulna
Abductor
Extensor pollicis longus tendon
First dorsal interosseous muscle
Tendons of extensor digitorum muscle
Avulsion of distal interphalangeal joint (Mallet finger)
TENDON AVULSION INJURIES
Avulsion at proximal interphalangeal joint (Boutonniere deformity)
Avulsion at metacarpo-phalangeal joint (skier's thumb)
Colles' fracture
Smith's fracture
Bennett's fracture
FRACTURES
Boxer's fracture
Phalangeal fracture
Metacarpal fracture fixations
Scaphoid fracture

HIP AND KNEE
Ilio-femoral ligament
Greater trochanter
Lesser trochanter
Ischium
Ischio-femoral ligament
POSTERIOR VIEW OF HIP JOINT
Ilio-femoral ligament
Anterior superior iliac spine
Anterior inferior iliac spine
Lesser trochanter
ANTERIOR VIEW OF HIP JOINT
Ilium
Lumbar vertebrae
Sacrum
Pubis
Femur
Lateral femoral condyle
Medial femoral condyle
Lateral meniscus
Medial meniscus
Anterior cruciate ligament
Fibula
Tibia
Acetabulum
Fibrous capsule
Intracapsular fracture line
Intertrochanteric fracture line
Ligament of head of femur
Neck of femur
Compression hip screw stabilizing hip fractures
FRONTAL SECTION
Plastic socket (x-section)
Cement fixation
Femoral stem of prosthesis
Cement
TOTAL HIP REPLACEMENT
Anterior superior iliac spine
Greater trochanter
Head of femur
Ilio-femoral ligament
Acetabulum
Ligament
Ligament of head of femur
Lesser trochanter
LATERAL VIEW OF HIP JOINT (Dislocated)
Head of femur
Neck
Ligament
Capsule
Necrosis
INTERTROCHANTERIC FRACTURE
INTRACAPSULAR FRACTURE
DISLOCATION
BLOOD SUPPLY TO HEAD OF FEMUR
Blood flow to the head of the femur is supplied by arteries coursing through the capsule and ligament. When a fracture or dislocation tears the capsule and ligament, the head of the femur loses its blood supply and deteriorates (avascular necrosis).
Lateral meniscus
Posterior cruciate ligament
Tibial collateral ligament
Medial meniscus
Anterior cruciate ligament
Posterior tears of lateral meniscus
Posterior tear of medial meniscus
"Bucket handle" tear
Anterior cruciate ligament
Posterior cruciate ligament
Medial meniscus
Meniscus profile
Tibia
Fibula
TIBIAL PLATEAU
Medial femoral condyle
Meniscus cutter
Cannula for fluid
Meniscus cutter
Medial tibial plateau
Medial meniscus tear
Arthroscope
ARTHROSCOPY
The arthroscope, attached to a video camera, is inserted into the fluid-filled knee. Next, the surgeon inserts small instruments through a second incision. Guided by the image on the TV monitor, he explores the menisci, and, if need be, removes or trims damaged tissue.
MENISCUS AND LIGAMENT TEARS
When the knee bends, the menisci stretch to accommodate the movement. When the knee bends and twists, the menisci may overstretch and tear. The medial meniscus is especially vulnerable to tearing because it is anchored to the tibial collateral ligament, and so has less mobility than the lateral meniscus.
A direct blow to the knee may tear the supporting ligaments. The direction of impact determines which ligaments are injured.
Twisting
Twisting
Anterior cruciate ligament tear
Posterior cruciate ligament tear
"Unhappy triad"
① Torn tibial collateral ligament
② Medial meniscus tear
③ Anterior cruciate ligament tear
Fibular collateral ligament tear
Tibial collateral ligament tear
Tackling can result in an "Unhappy triad"
Quadriceps tendon
Articular surface of patella
Anterior cruciate ligament
Patellar tendon
Lateral meniscus
Fibular collateral ligament
ANTERO-LATERAL VIEW OF KNEE
Tibial collateral ligament
Medial meniscus
Medial femoral condyle
Posterior cruciate ligament
POSTERO-MEDIAL VIEW OF KNEE
POSTERIOR VIEW OF KNEE
Patellar prosthesis
Femoral condylar prosthesis
Cement
Tibial prosthesis
TOTAL KNEE REPLACEMENT
©1986 Anatomical Chart Co., Skokie, Illinois

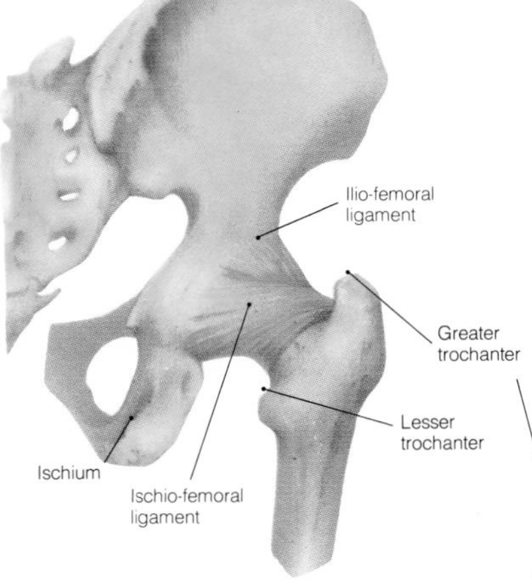

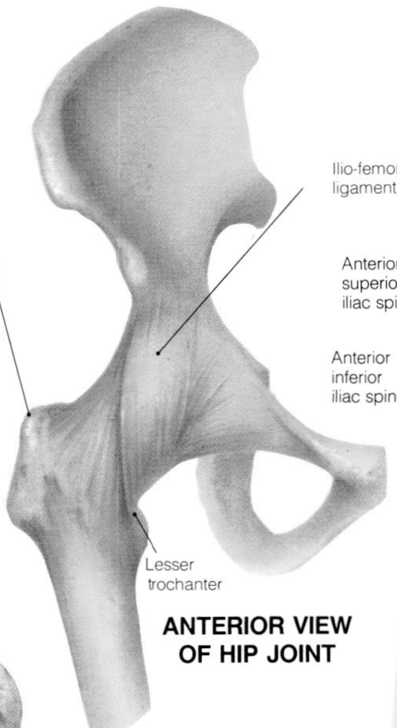

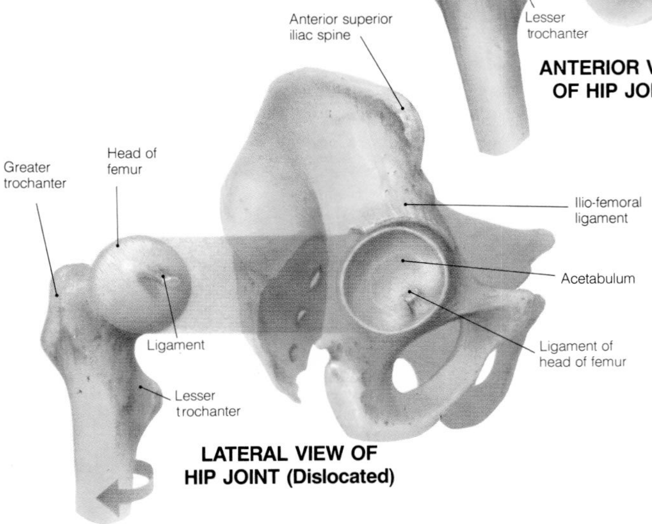

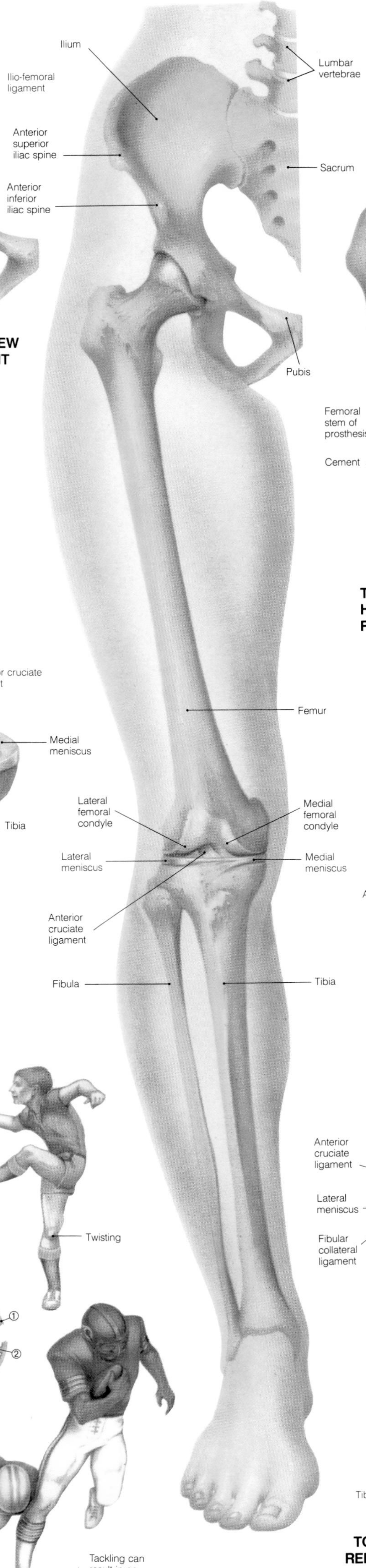

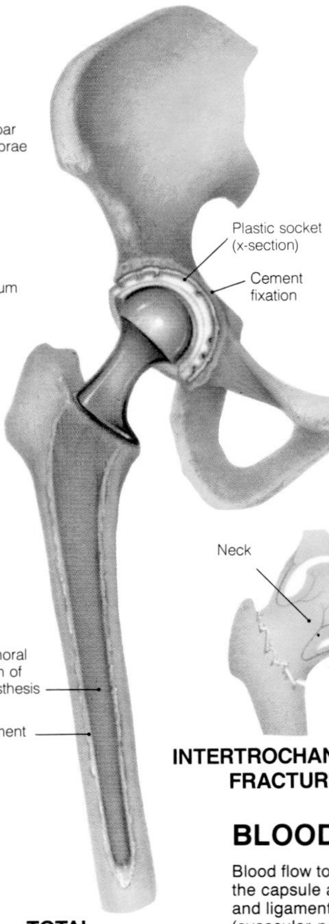

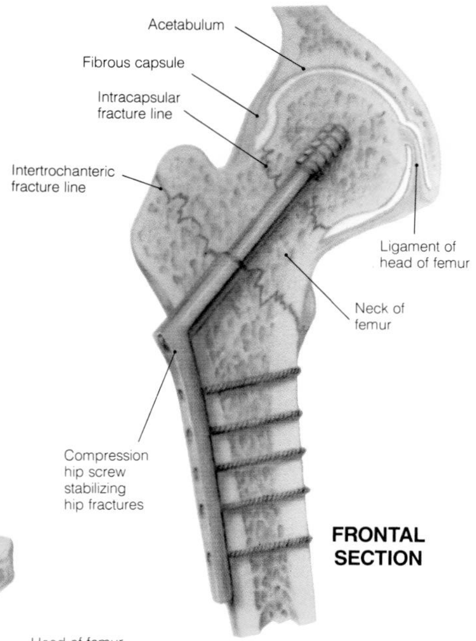

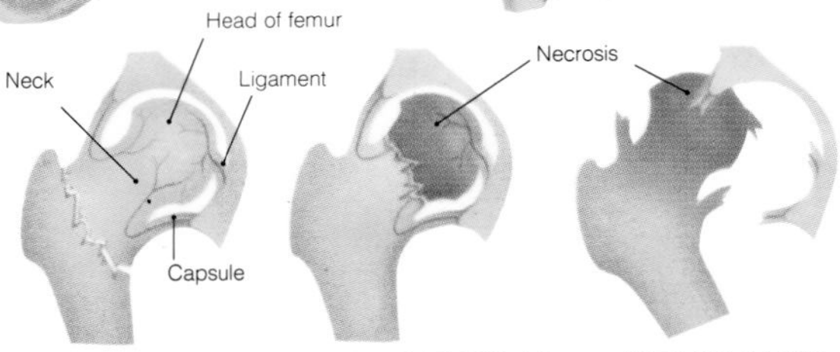

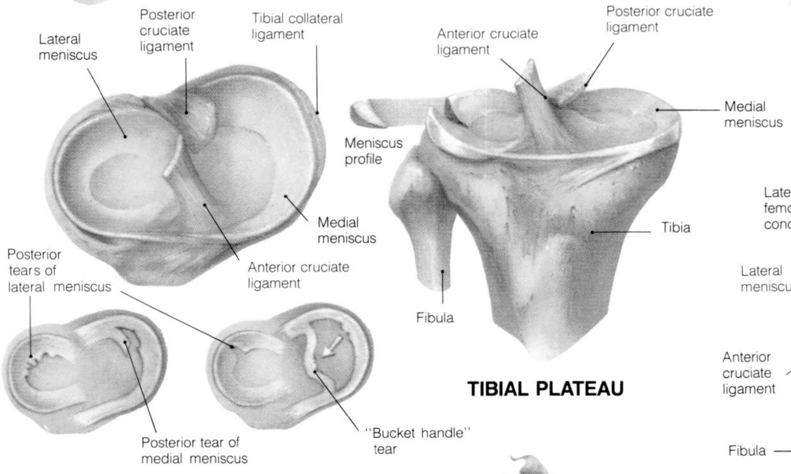

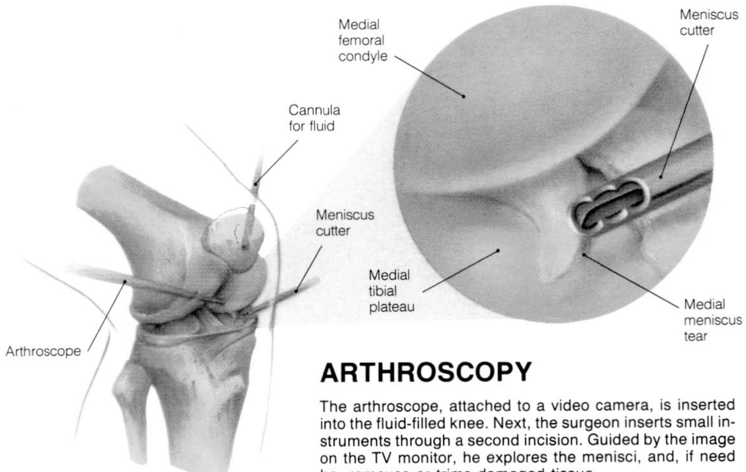

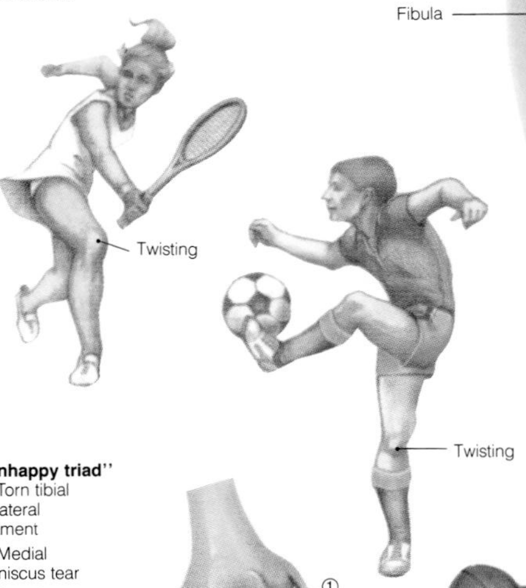

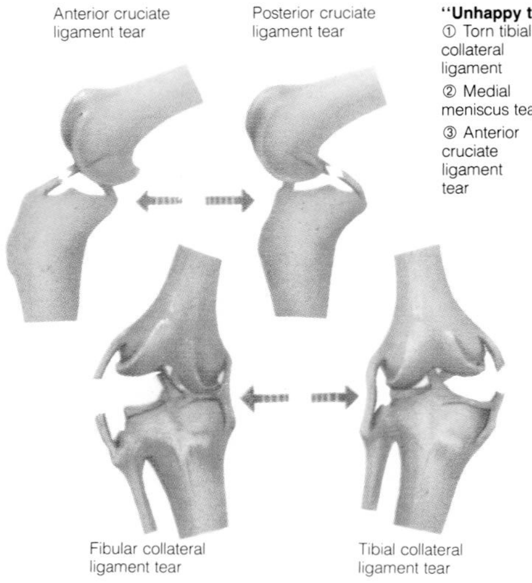

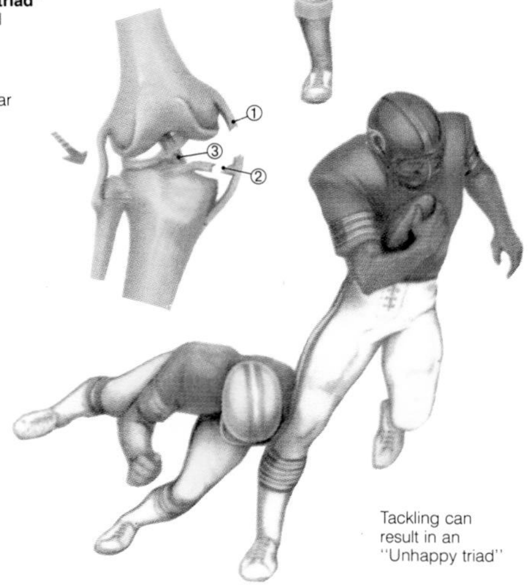

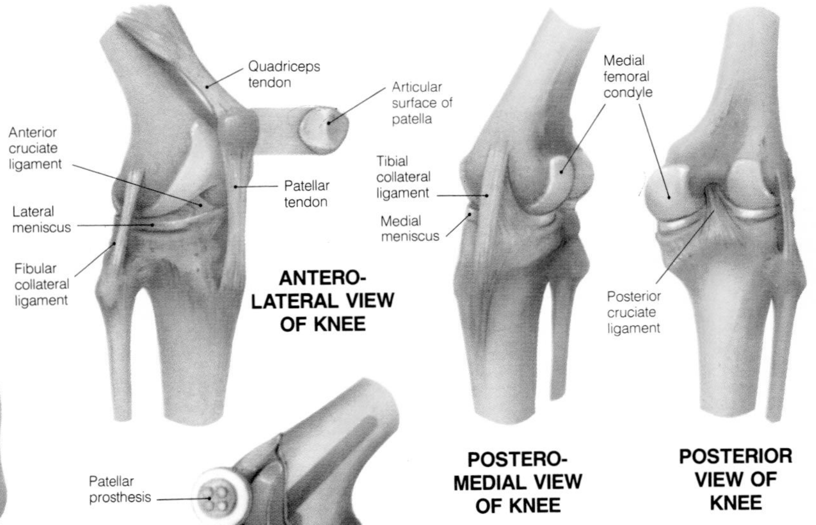

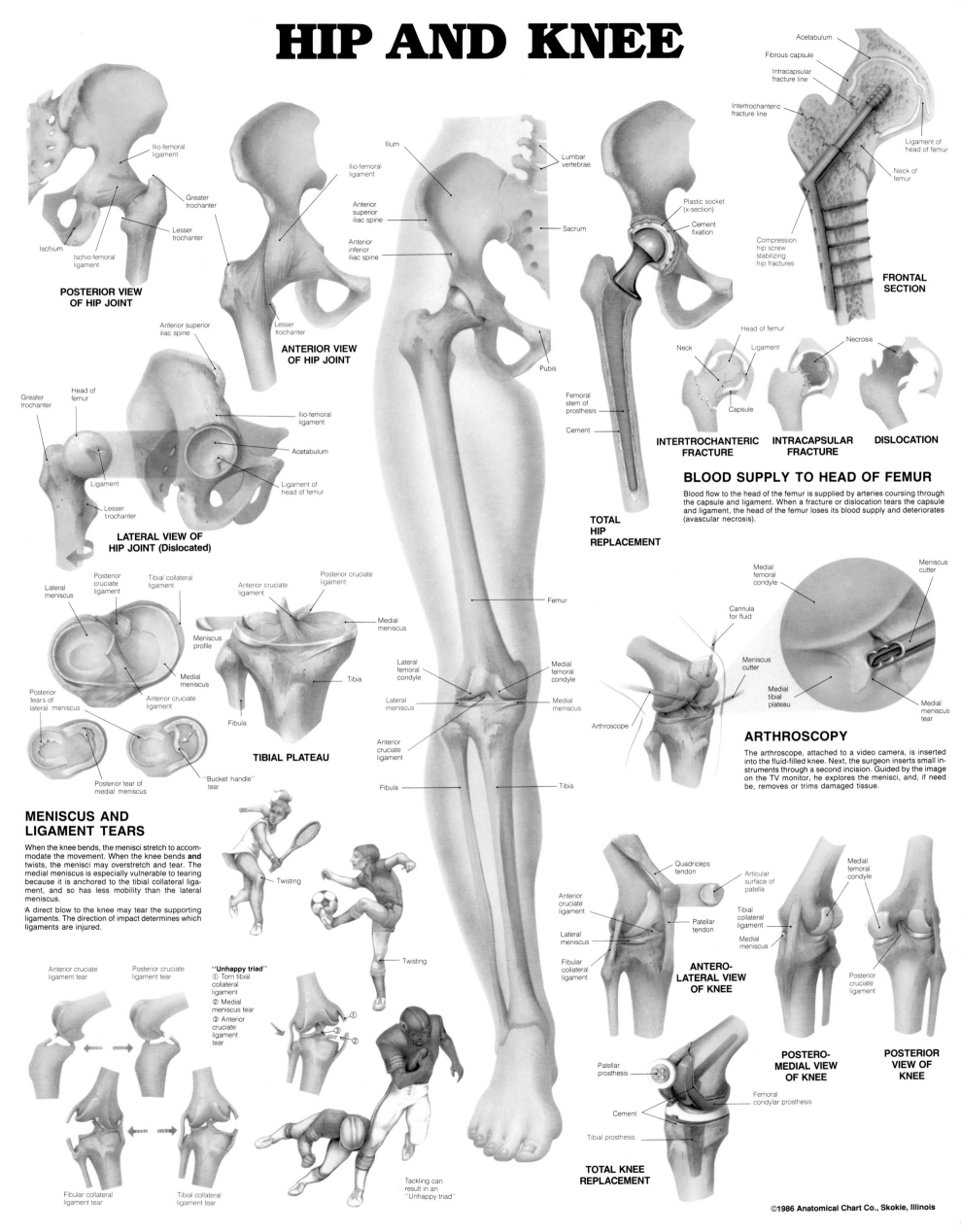

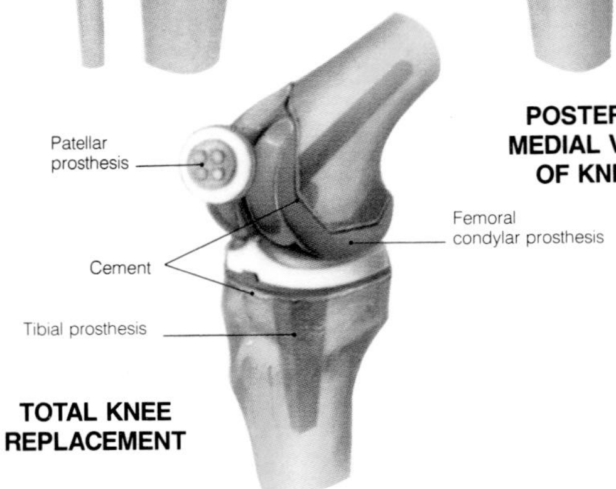

FOOT AND ANKLE

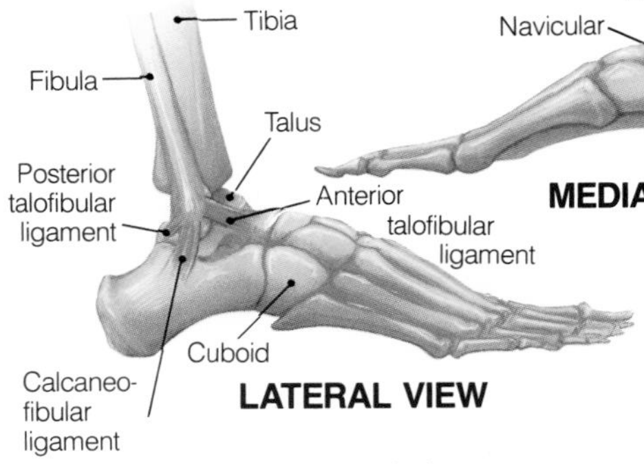

LATERAL VIEW

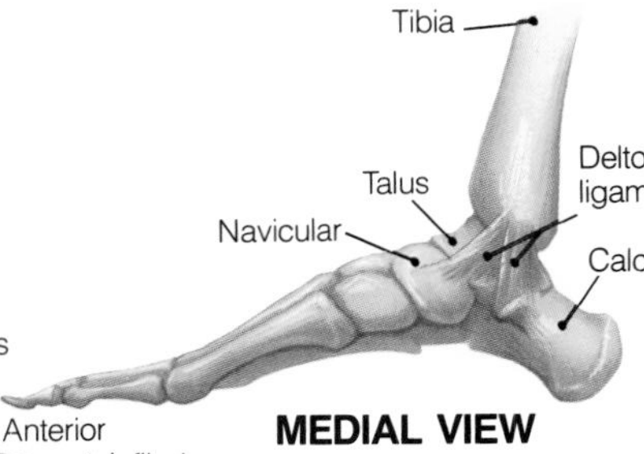

MEDIAL VIEW

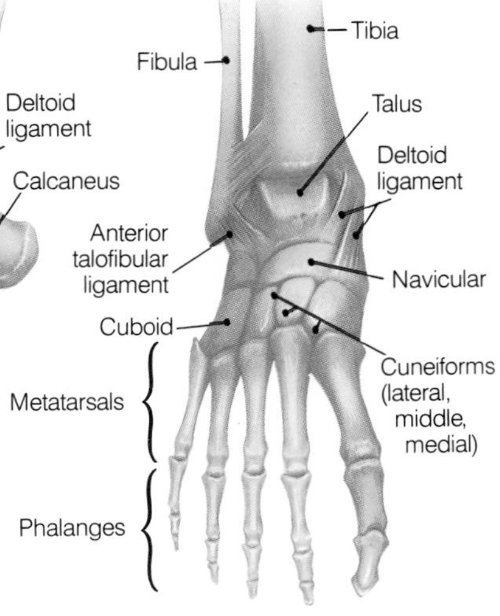

FRONTAL VIEW

THE ANKLE JOINT

The ankle joint consists of three bones, the tibia (medial malleolus,) the fibula (lateral malleolus,) and the talus. It is a hinged joint responsible for dorsiflexion (upward motion) and plantar flexion (downward motion) of the foot. Inversion (inward motion) and eversion (outward motion) of the foot take place in the joints below the talus. The bony elements of the ankle joint are held together by ligaments.

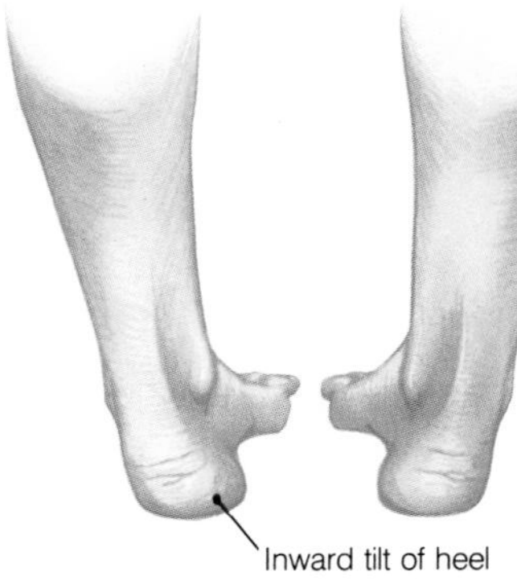

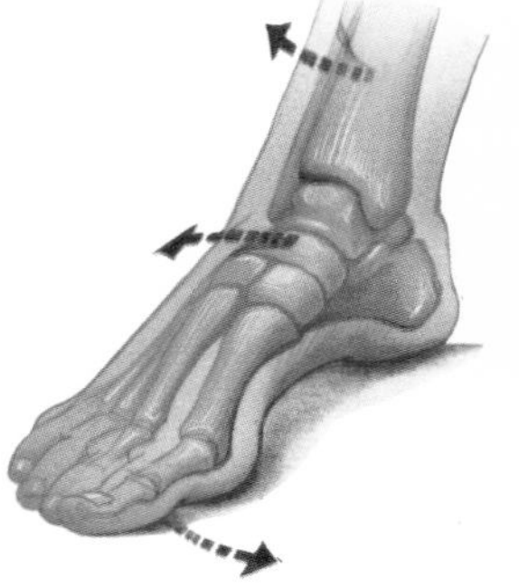

HIGH ARCH
(Supination)

PRONATION AND SUPINATION

Pronation and supination are complex motions around the joint beneath the ankle. Both the pronated (low arch, flat) foot and the supinated (high arch, cavus) foot may have associated symptoms that benefit from orthotics (supportive shoe inserts).

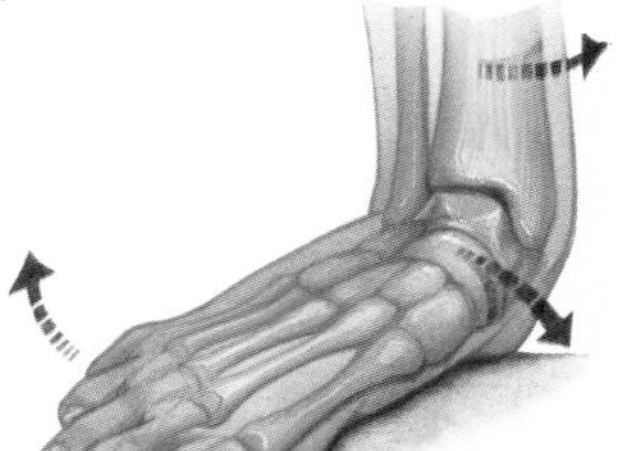

LOW ARCH
(Pronation)

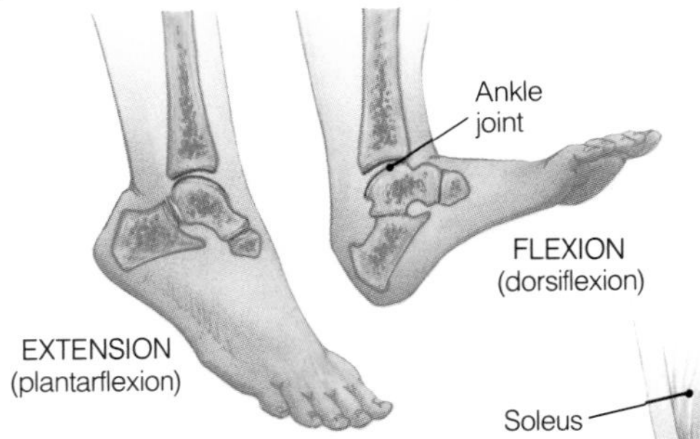

X-SECTION

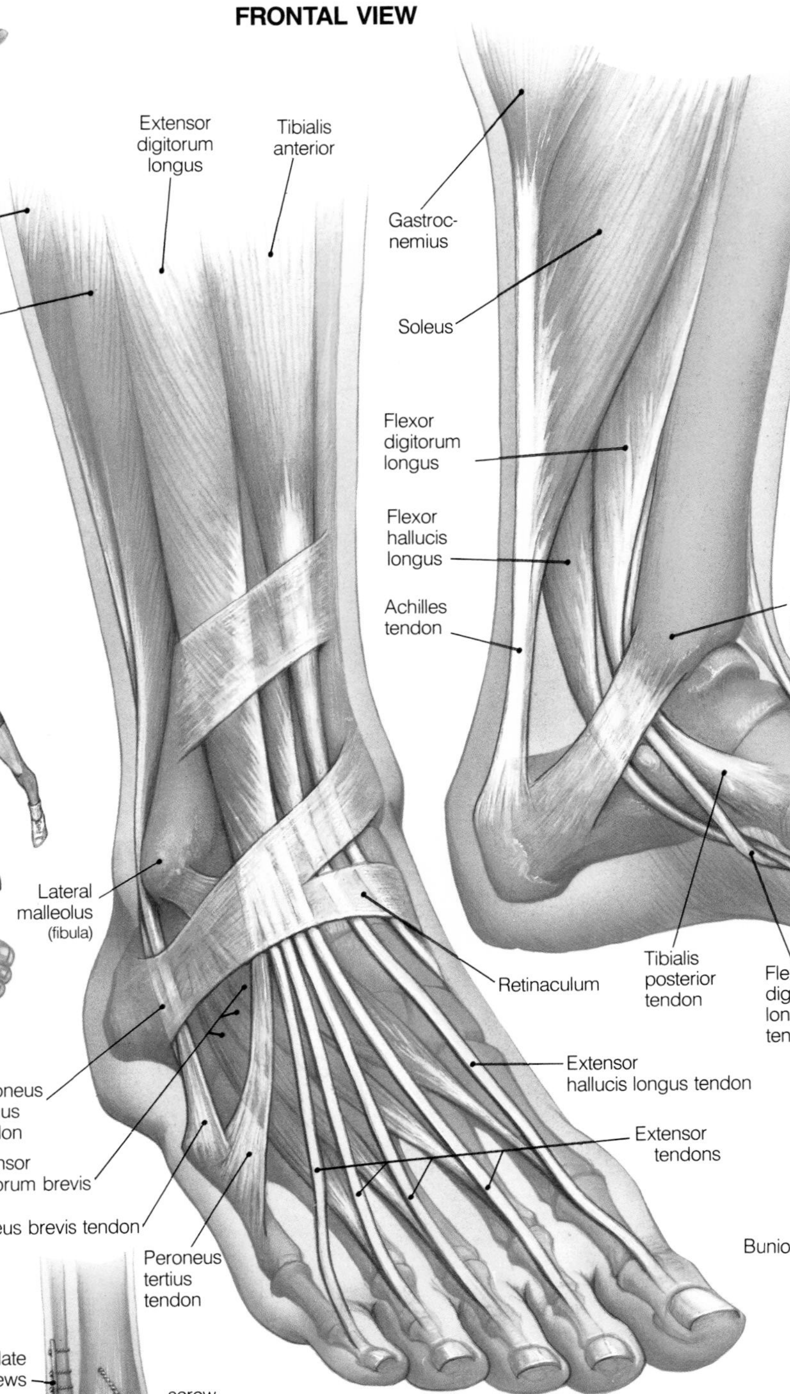

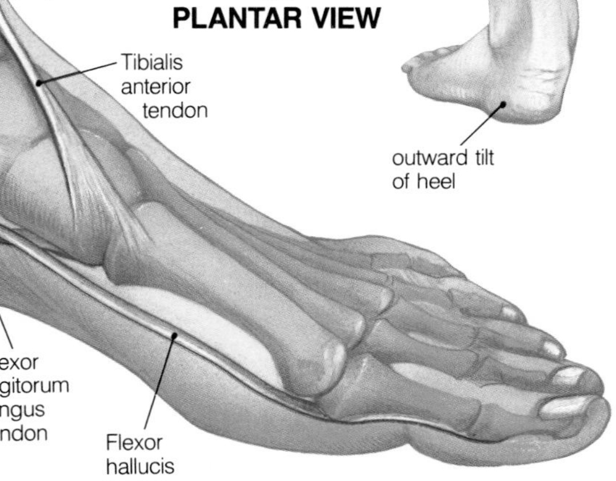

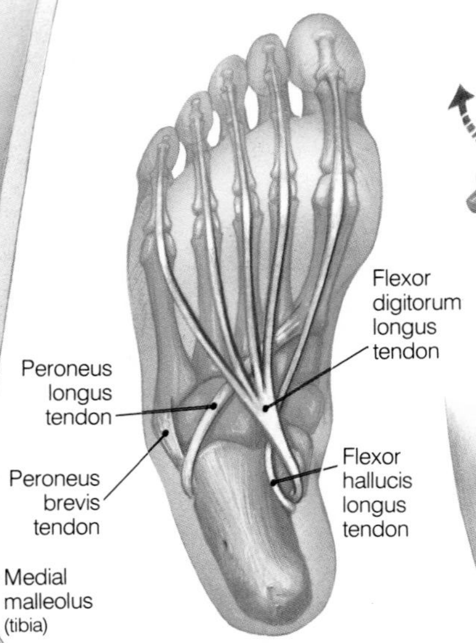

PLANTAR VIEW

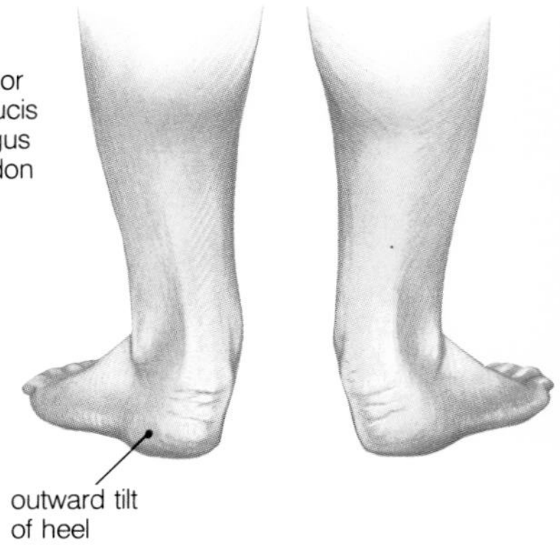

SPRAINS

Sprains are injuries to ligaments, those structures that attach one bone to another. The degree of severity is directly related to the degree of involvement of the three bands of the lateral ligament complex (anterior talofibular, posterior talofibular, calcaneal fibular.)

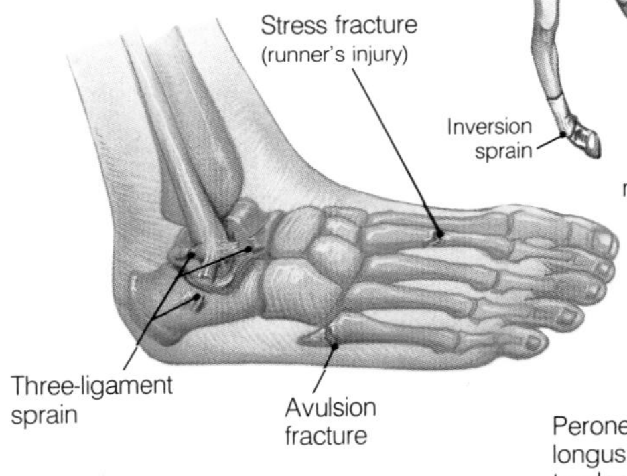

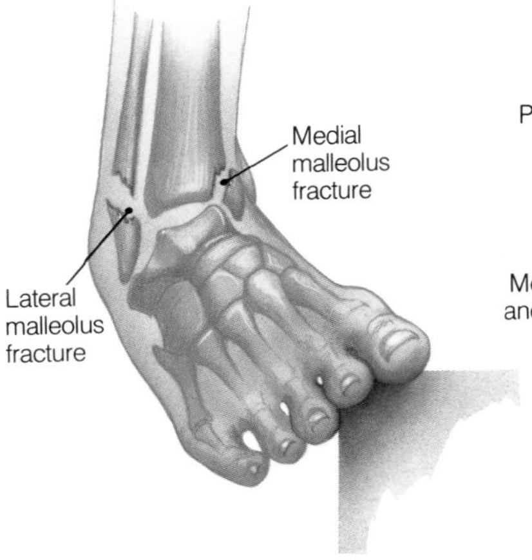

FRACTURES

Fractures are breaks in the bony architecture of the ankle that result from severe stresses. Displacement often requires surgical intervention to best restore normal anatomic relationships.

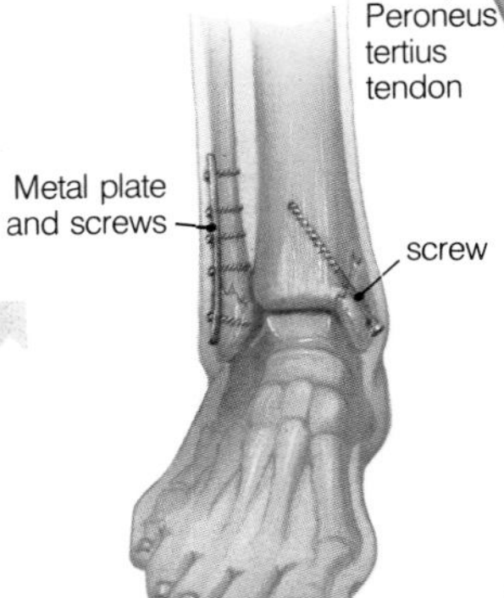

FRACTURE FIXATION

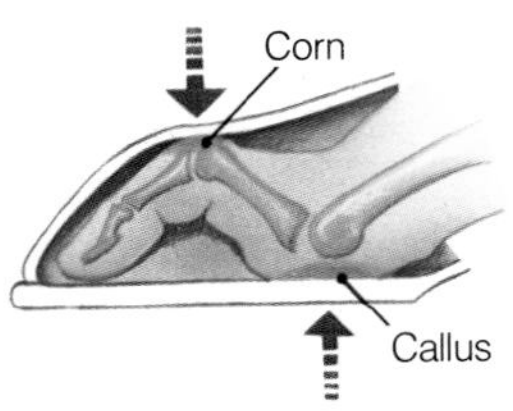

HAMMERTOE

(Common flexion deformity of the lesser toes)

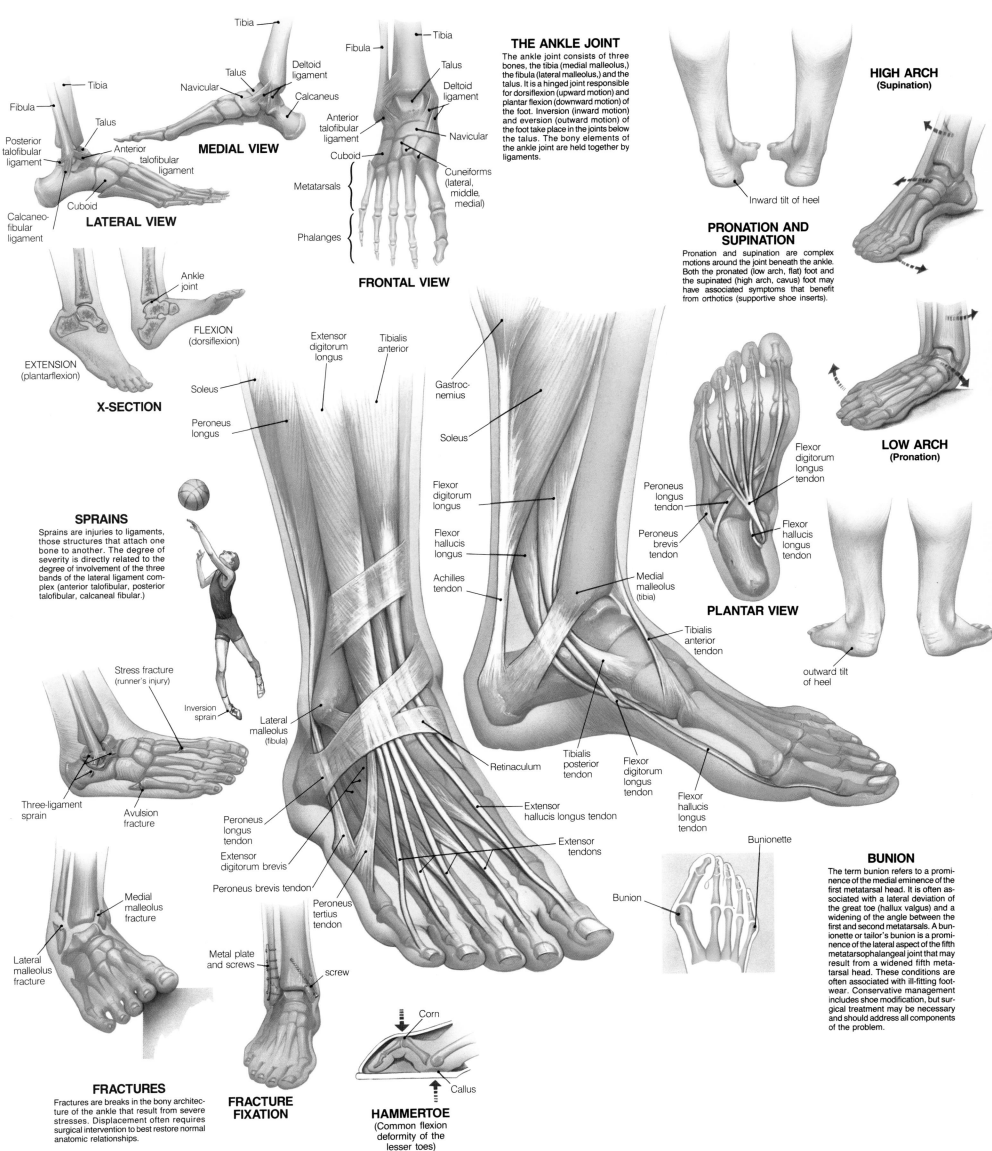

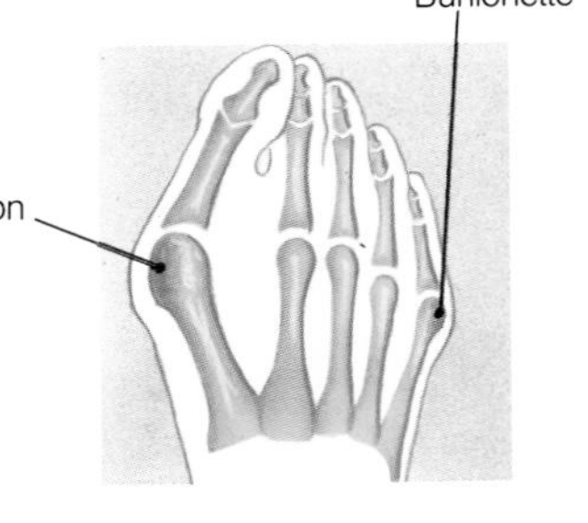

BUNION

The term bunion refers to a prominence of the medial eminence of the first metatarsal head. It is often associated with a lateral deviation of the great toe (hallux valgus) and a widening of the angle between the first and second metatarsals. A bunionette or tailor's bunion is a prominence of the lateral aspect of the fifth metatarsophalangeal joint that may result from a widened fifth metatarsal head. These conditions are often associated with ill-fitting footwear. Conservative management includes shoe modification, but surgical treatment may be necessary and should address all components of the problem.

THE TEETH

Anatomy of a Tooth

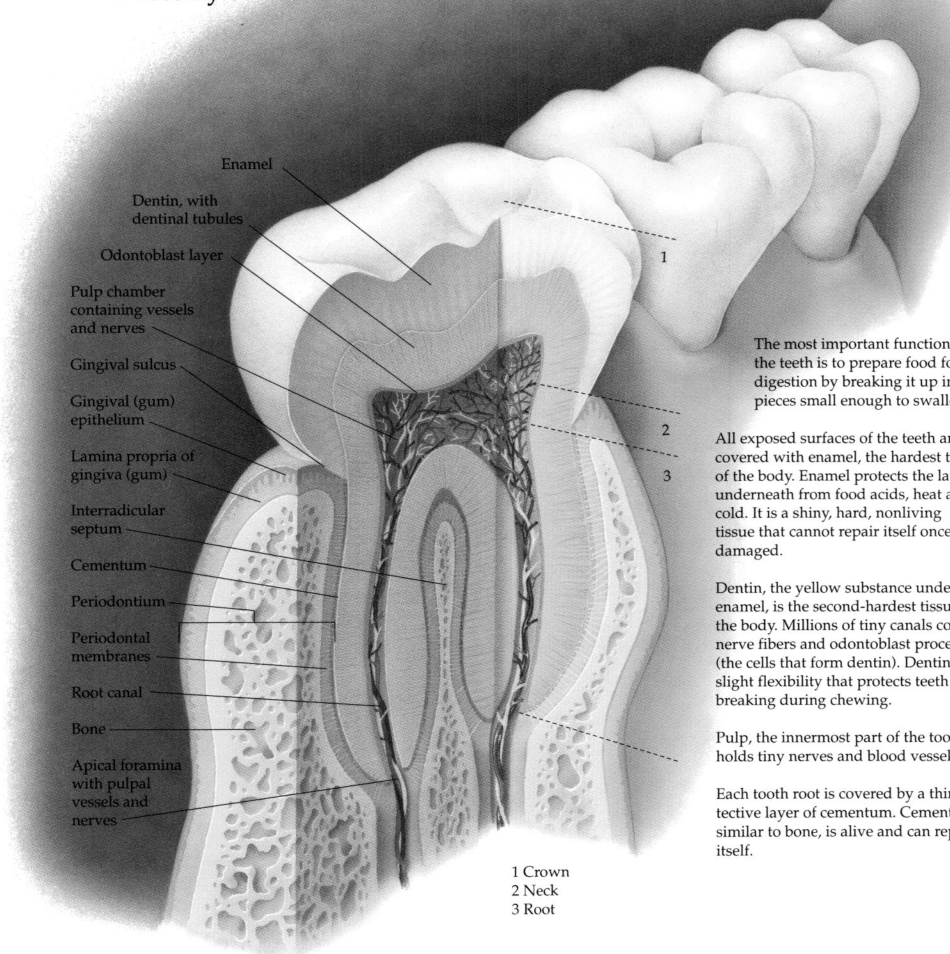

1 Crown
2 Neck
3 Root

The most important function of the teeth is to prepare food for digestion by breaking it up into pieces small enough to swallow.

All exposed surfaces of the teeth are covered with enamel, the hardest tissue of the body. Enamel protects the layers underneath from food acids, heat and cold. It is a shiny, hard, nonliving tissue that cannot repair itself once damaged.

Dentin, the yellow substance under the enamel, is the second-hardest tissue of the body. Millions of tiny canals contain nerve fibers and odontoblast processes (the cells that form dentin). Dentin has a slight flexibility that protects teeth from breaking during chewing.

Pulp, the innermost part of the tooth, holds tiny nerves and blood vessels.

Each tooth root is covered by a thin protective layer of cementum. Cementum is similar to bone, is alive and can repair itself.

Primary Teeth

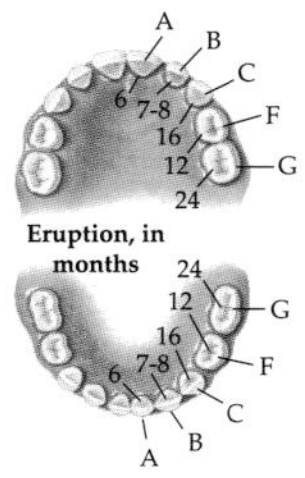

Permanent Teeth

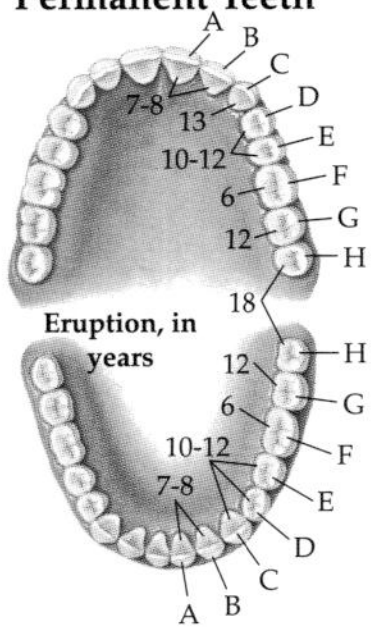

Oral Cavity

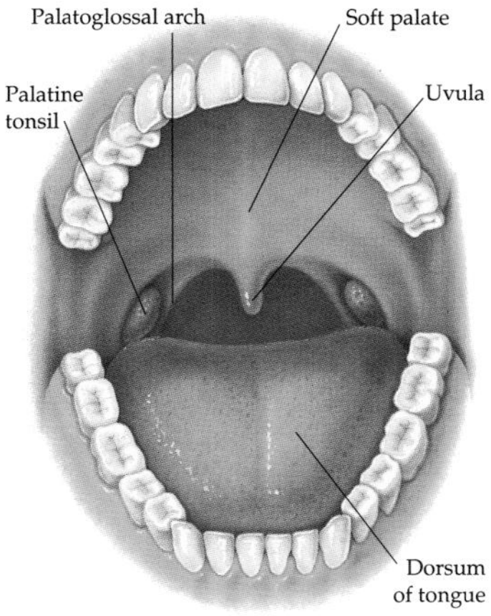

A Central incisor
B Lateral incisor
C Canine
D First premolar
E Second premolar
F First molar
G Second molar
H Third molar

Function of Teeth

Incisor: Acts like scissors; grasps and cuts off food pieces.

Cuspid: Has a single, very long, sharp cusp; tears and shreds food.

Bicuspid: Has two pointed projections; tears, shreds, crushes food.

Molar: Strongest, most useful type of tooth; grinds food into tiny pieces.

Innervation and Blood Supply

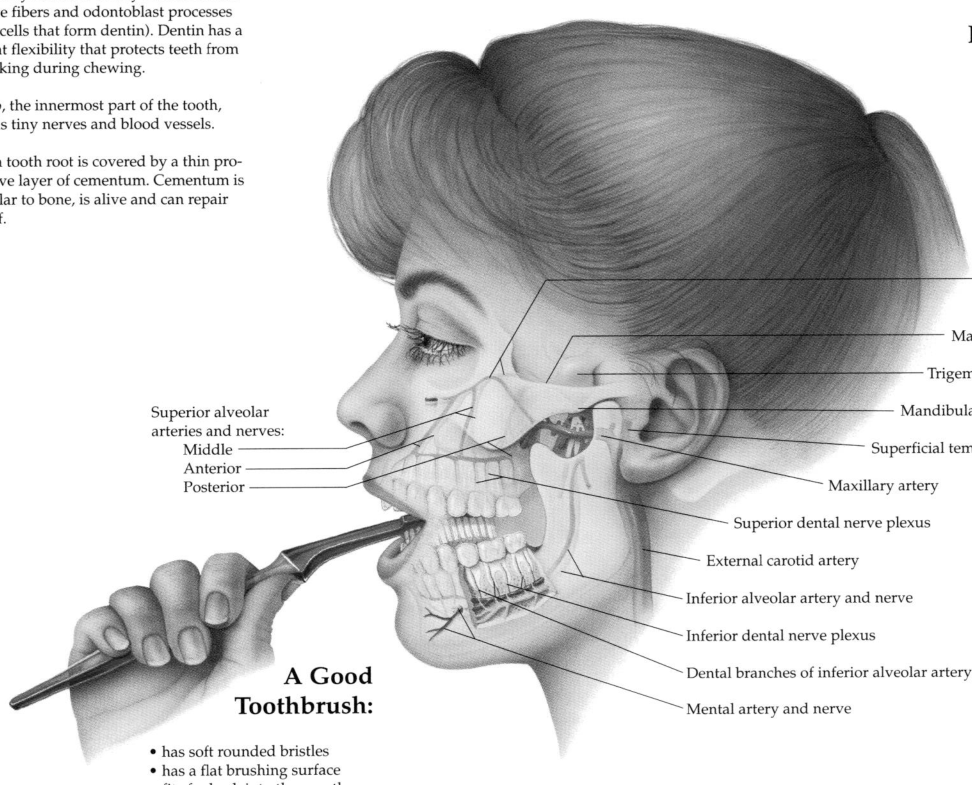

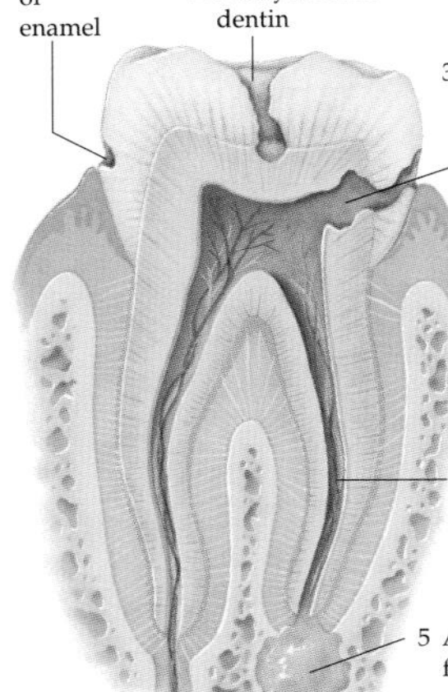

Tooth Decay (Caries)

Tooth decay, called caries, is a softening of the surface of the tooth. Caries is caused by **plaque**, a sticky film full of bacteria that convert sugar and starch in food into acid that dissolves tooth enamel. Plaque builds up around the teeth and gums, and in the pits and grooves of the molars. Decay from food acids, if not stopped, can travel through the enamel down into the dentin. Decay reaching the pulp can cause throbbing pain, a **toothache.**

Parts of the teeth where decay most often occurs are the grooves of the molars, contact points where adjacent teeth meet, and at the necks of teeth above the sulci.

A Good Toothbrush:

- has soft rounded bristles
- has a flat brushing surface
- fits far back into the mouth
- is replaced when bristles start to bend out of line

Tooth Repair

Fillings

There are several materials used to fill cavities. If the filling will show, a tooth-colored resin or porcelain is used. If not, gold casting or a mixture of silver, mercury, and other metals called **amalgam** is commonly used.

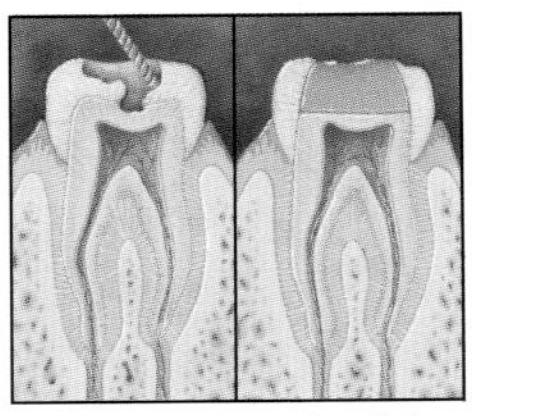

*Area of decay is removed by drilling with small cutters called **burs**.*

Resulting hole is cleaned, shaped and filled.

Root Canal Therapy

Tooth decay, trauma, and gum disease can injure or cause an infection in the pulp of a tooth, leading to pain, sensitivity to hot and cold, or no symptoms. Often, root canal therapy can save the tooth.

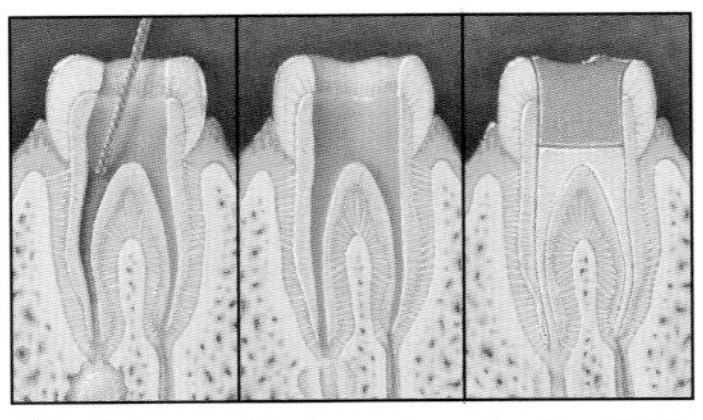

All pulp is removed from the chamber and root canals of the tooth.

Empty canal is cleaned, shaped, and sterilized with antiseptic.

*Canals are filled with a resin called **gutta percha**, and the access channel is filled.*

Tooth Care

General Rules of Tooth Brushing:

- brush teeth in a certain order which will become routine
- brush every reachable surface of each tooth
- brush at least twice a day, after meals. It is especially important to brush before going to sleep
- brush with flouride toothpaste

Brushing

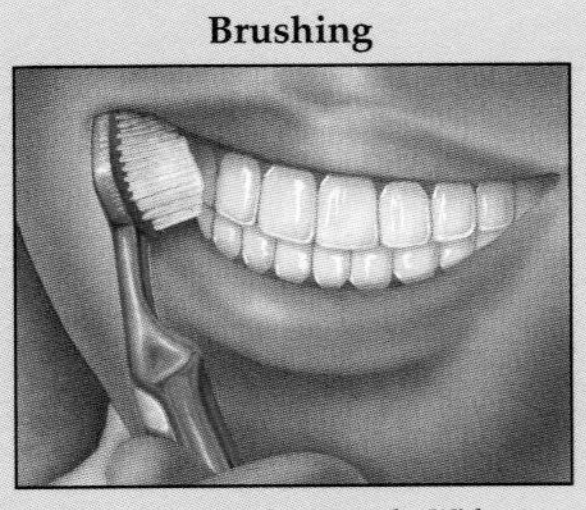

Hold brush at a 45 degree angle. With light pressure, brush back and forth with short strokes on teeth and gums. Rinse thoroughly. Correct brushing should take 2-4 minutes.

Flossing

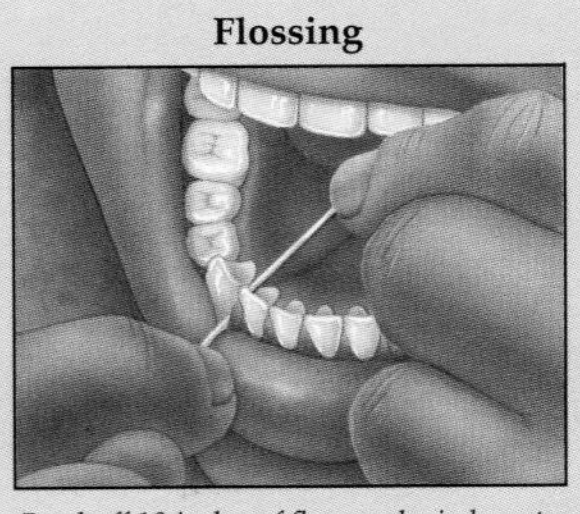

Break off 18 inches of floss, and wind most of it around the middle finger of one hand. Keeping one inch taut between your fingers, guide floss between teeth to the gum. Scrape along the side of each tooth, and repeat several times. Move to next tooth. After flossing, rinse mouth thoroughly.

Flossing is an essential component of tooth care because it reaches between teeth and beneath the gum line, where brush bristles cannot. If food particles remain in these places, plaque buildup will start. Ideally, flossing should be done after every meal, but before sleeping is most important.

Braces

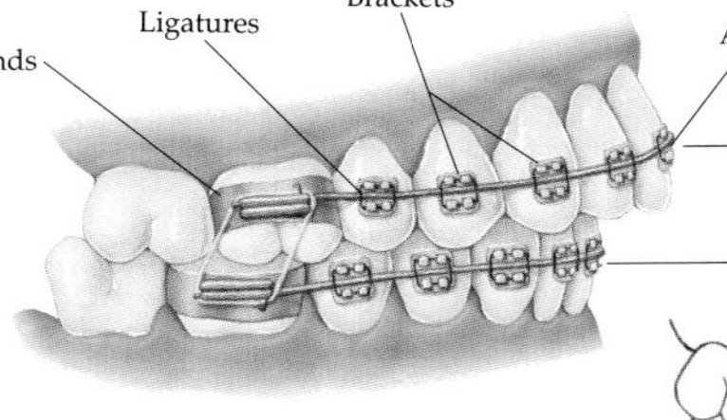

Proper alignment of the teeth is necessary to maintain healthy teeth and gums. It also enhances cosmetic appearance. In a normal bite, the lower first molars are slightly forward of the uppers, and the lower front teeth rest lightly against the backs of the upper ones. There are many types of imperfect bites, called **malocclusions**. One example is an **excessive overjet.** Orthodontic appliances are used to direct the growth of the jaws and movement of the teeth by applying steady but gentle pressure, working on the principal that bone is resorbed where pressure is applied. Brackets, bands and wires, called **braces**, allow precise control of applied pressure that is unsurpassed by other appliances.

Excessive Overjet: upper front teeth protrude, lower jaw closes too far behind upper one.

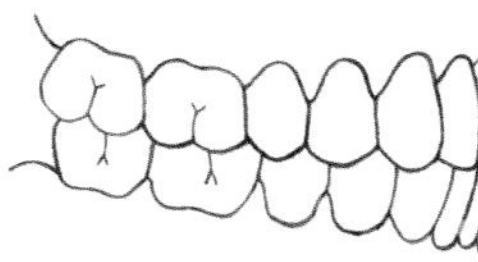

Normal Bite

Check-ups

The main purpose of modern dental treatment is prevention of tooth decay and gum disease. Even with proper brushing and flossing, it is still necessary to see your dentist once every six months to keep your teeth free of plaque and tartar. Your dentist will examine your teeth and gums, take routine x-rays to check for tooth decay, remove tartar build-up, and clean and polish your teeth with professional instruments.

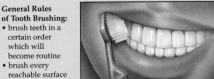

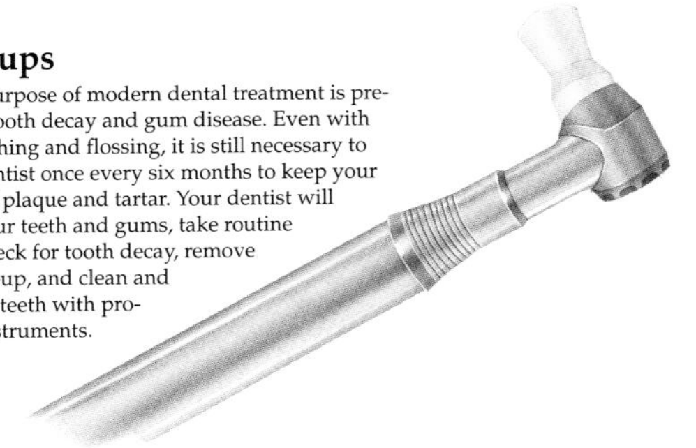

Healthy Eating

A healthy diet is essential to healthy teeth and gums. Developing teeth and the periodontal tissues especially need protein, vitamins A, C, and D, and calcium. It is also important to limit your intake of sweet foods like candy that will stick to your teeth and encourage bacterial growth that can lead to tooth decay. Vegetables, fruits and whole grain foods are great snacks because they are nutritious and won't stick to your teeth. If you eat sweets at all, eat them after a meal when saliva levels are high and can fight the acids produced by sugar.

PHARYNX & LARYNX

The Pharynx
(posterior view)

The pharynx is a muscular tube located above the esophagus and behind the nasal cavities, mouth and larynx.

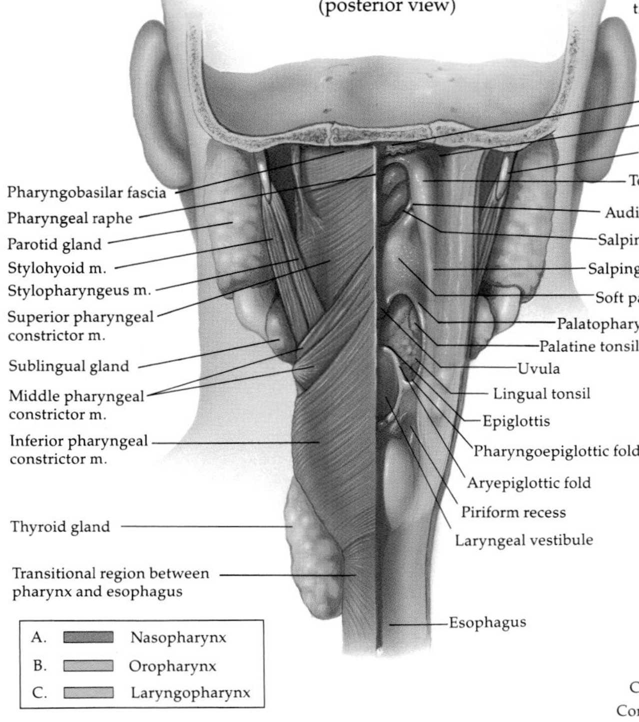

Sagittal section

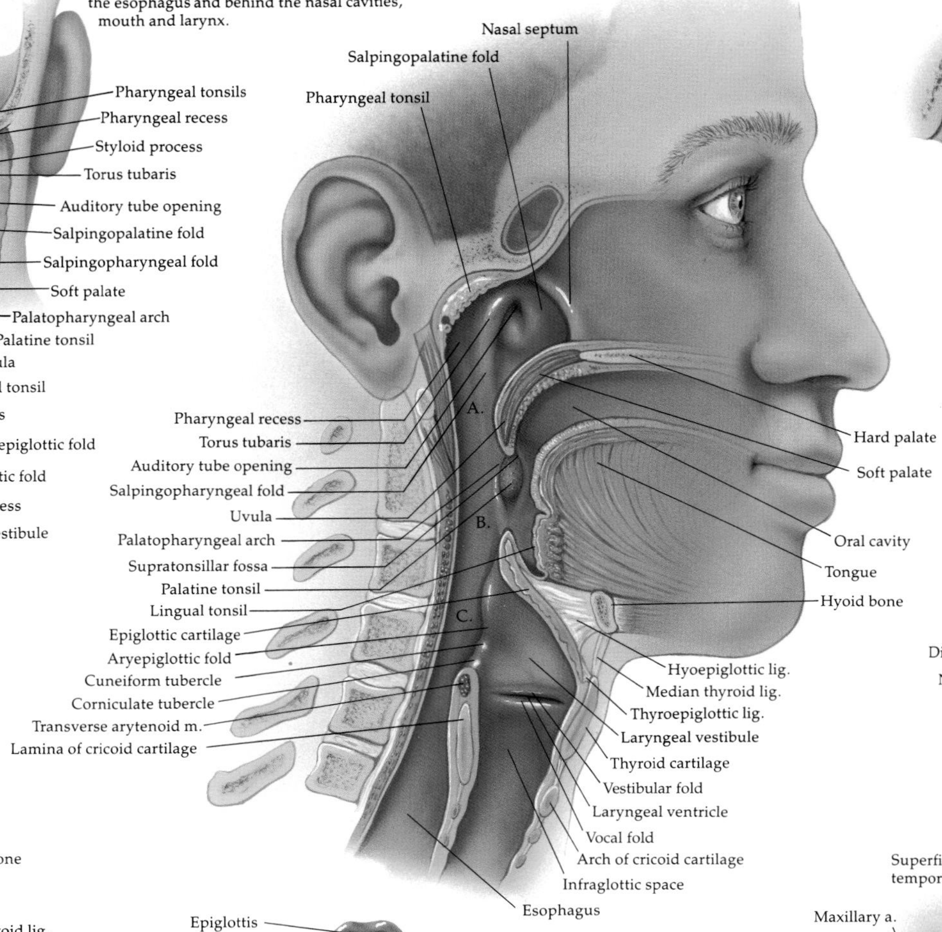

Tonsils

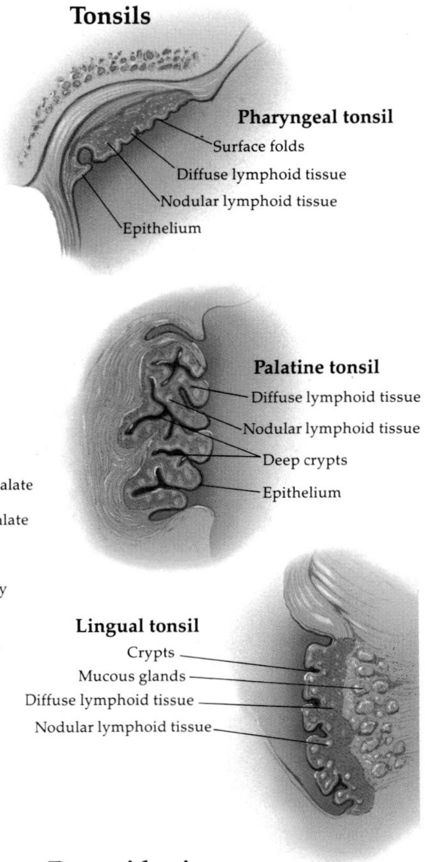

The Larynx

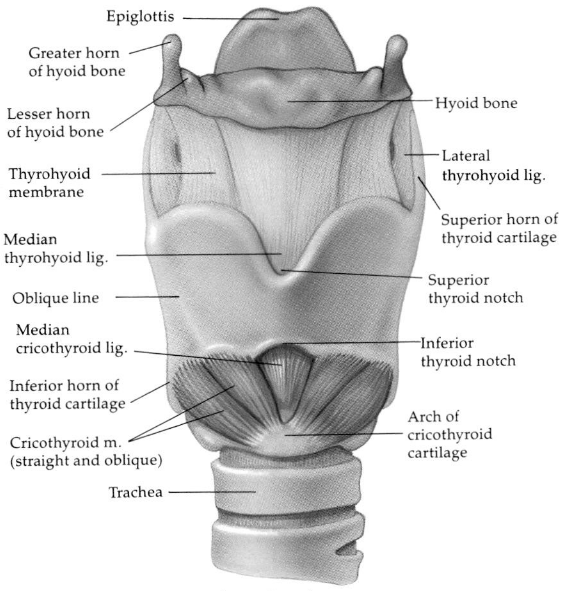

Anterior view

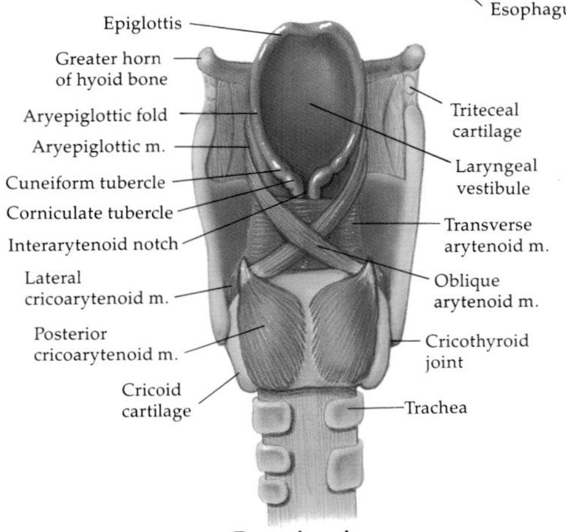

Posterior view

Deep side view

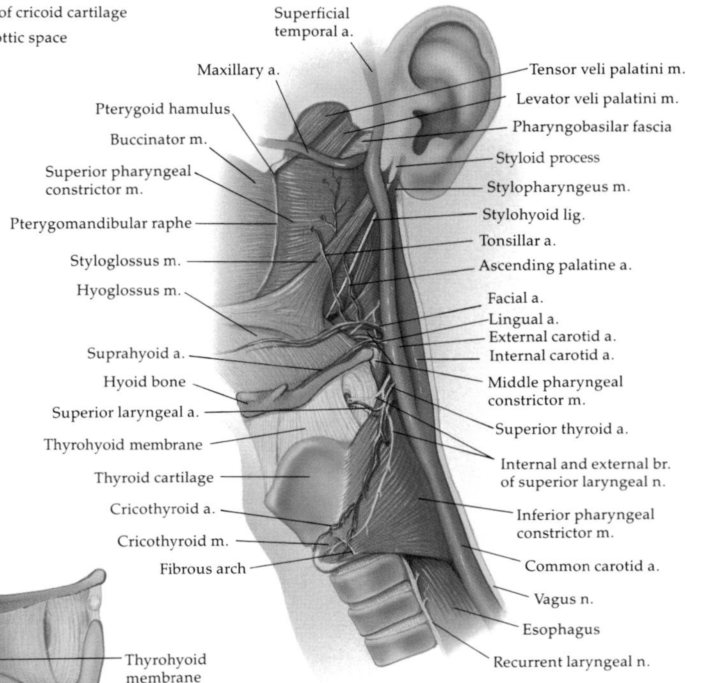

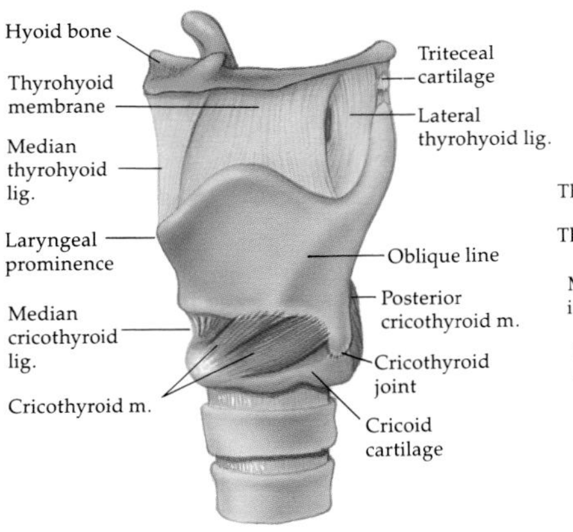

Side view

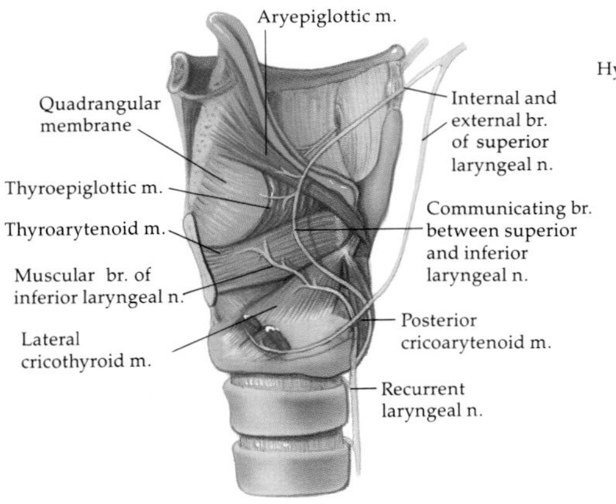

Cut-away side view

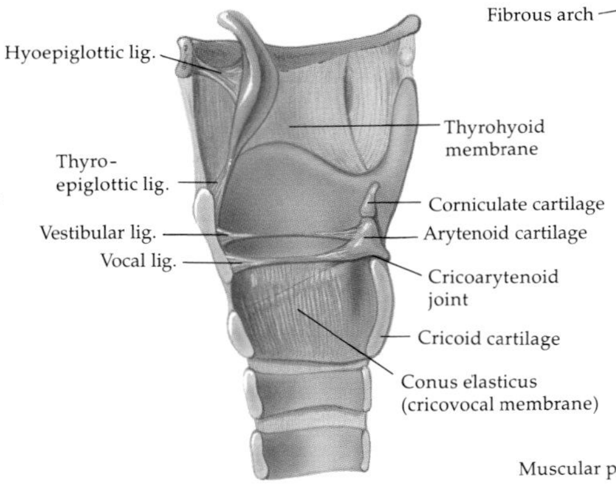

Sagittal section

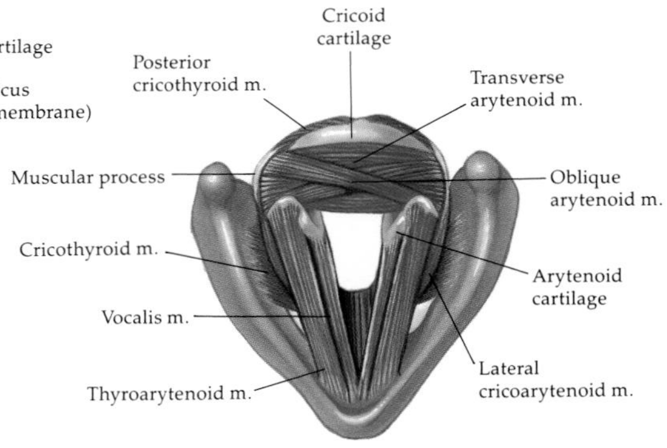

Top view
(epiglottis removed)

Laryngeal function
(mirror view)

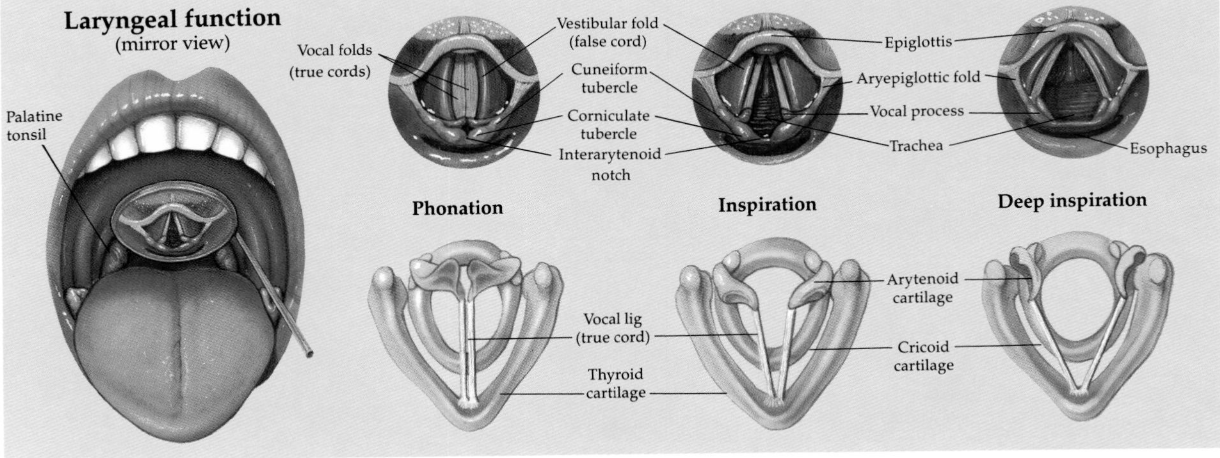

Medical consultant Dr. Malcolm H. Hast, Northwestern University Medical School, Chicago, Illinois.

PREGNANCY AND BIRTH

A. SPERMATOZOON

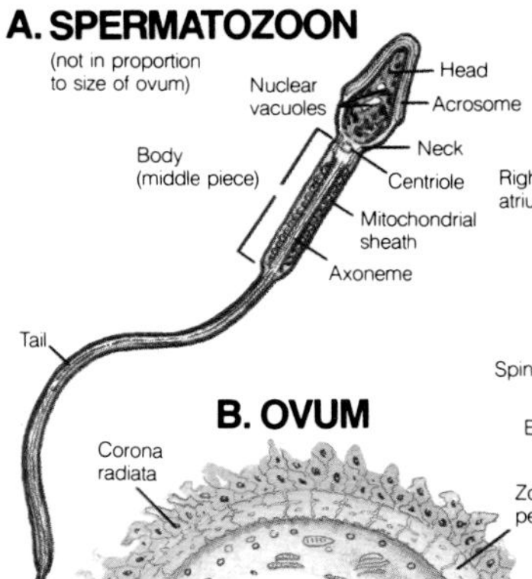

B. OVUM

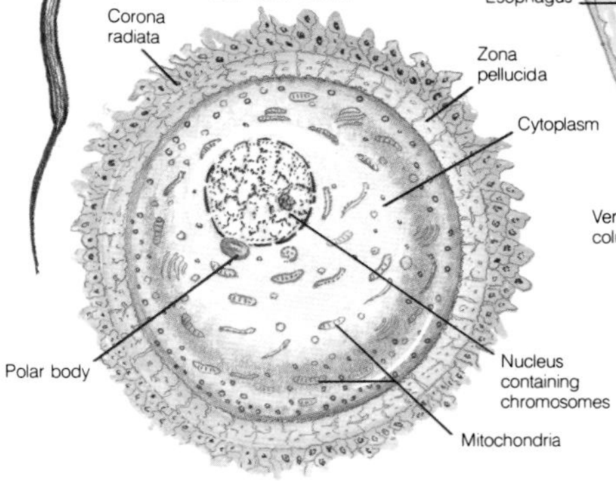

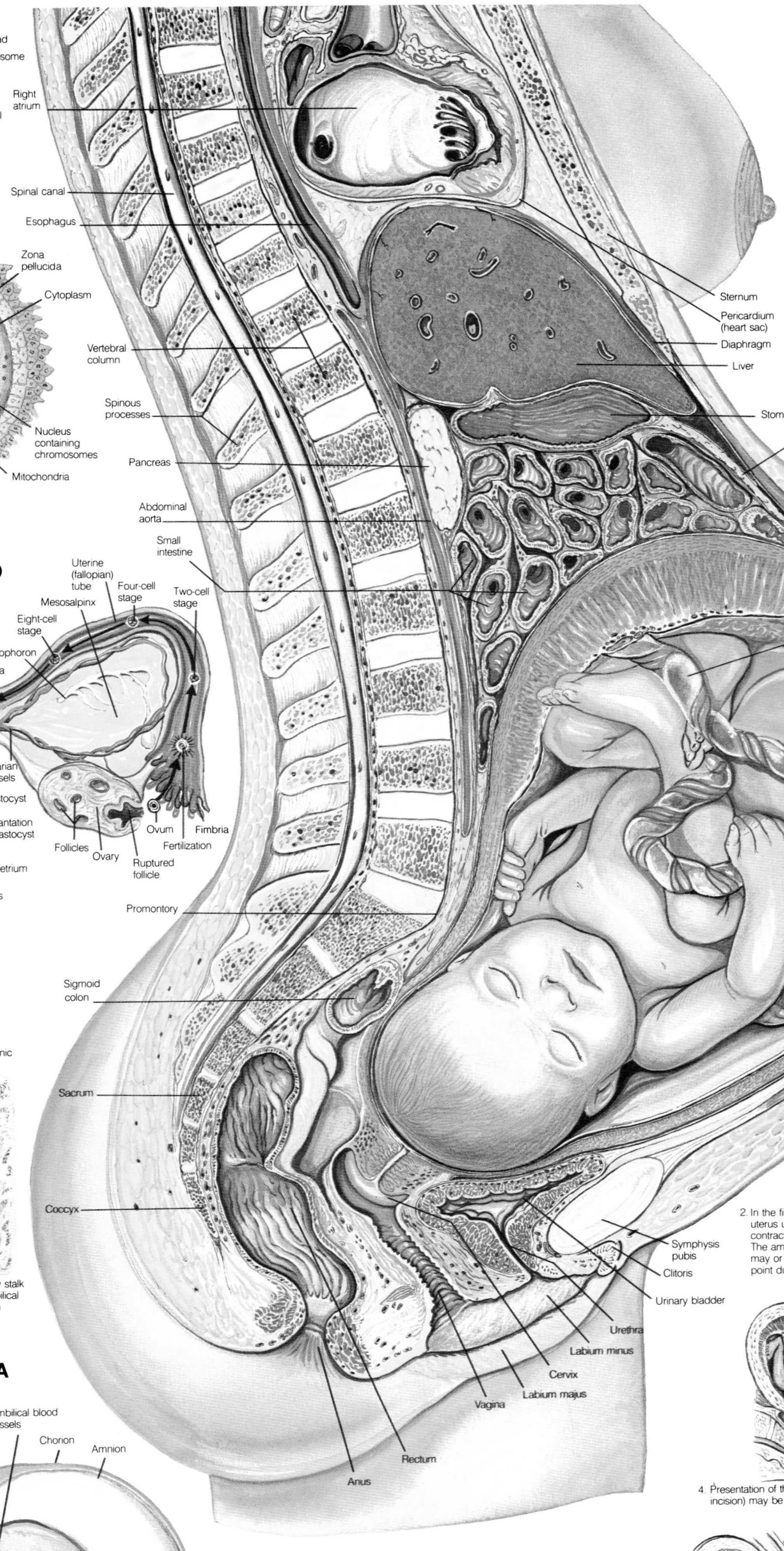

C. FERTILIZATION AND IMPLANTATION

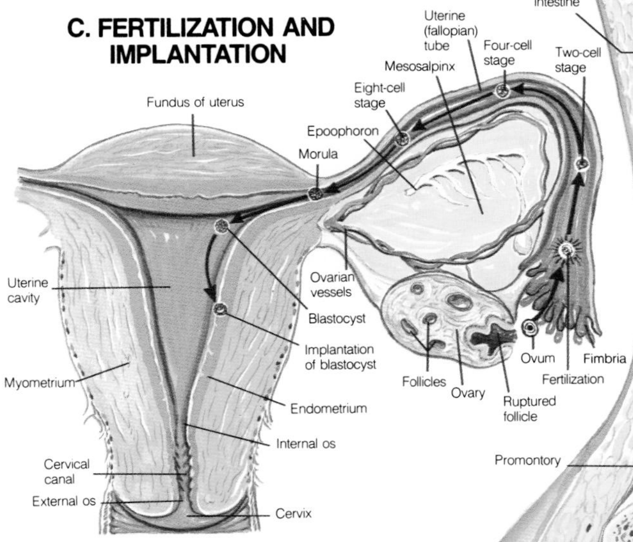

D. FORMATION OF FETAL MEMBRANES
(second week)

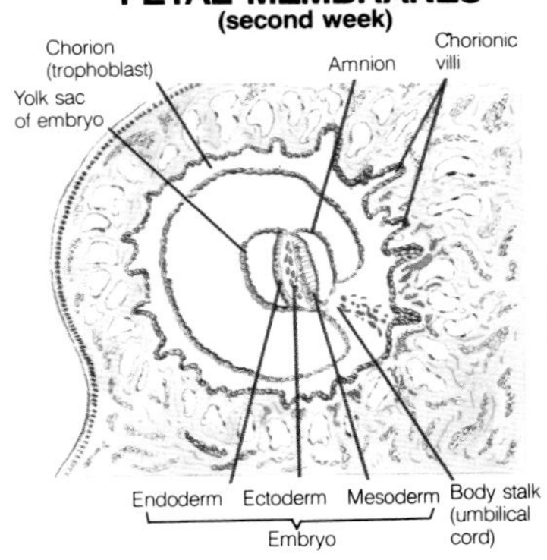

E. THE PLACENTA
(five weeks)

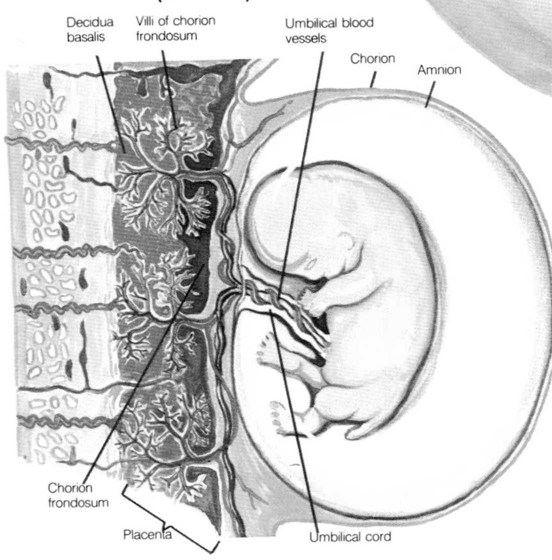

The placenta is the highly vascular structure that anchors the fetus to the mother's uterus. It is derived in part from both the developing embryo and maternal tissues. The placenta secretes both steroid hormones and glycoprotein hormones that have actions similar to those of some anterior pituitary hormones.

F. AMNIOCENTESIS

In this procedure, amniotic fluid, which contains cells sloughed from the baby's body, is withdrawn for examination, usually about the twelfth week of pregnancy. These cells can then be tested for certain genetic diseases and chromosomal abnormalities, such as Down syndrome. A newer procedure involves obtaining samples from the fetal placenta for examination of cells.

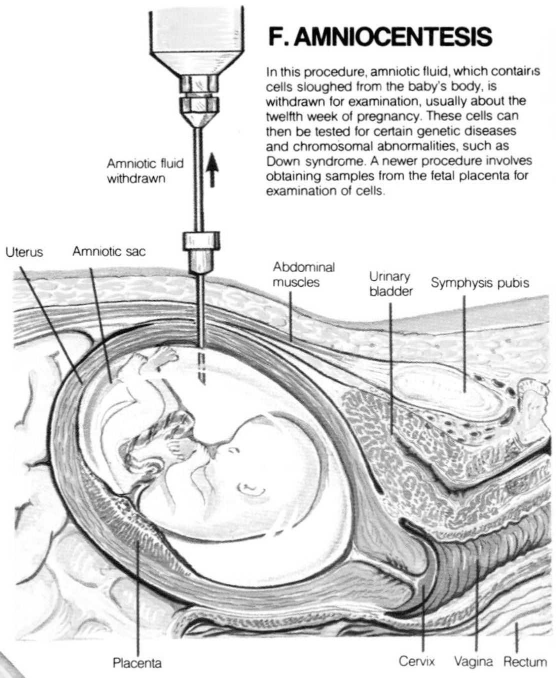

G. ULTRASOUND

Major structural abnormalities which may not be predictable from genetic analysis can often be detected by an image produced by sound vibrations reflecting from the interface of tissues with different densities.

H. LABOR AND BIRTH

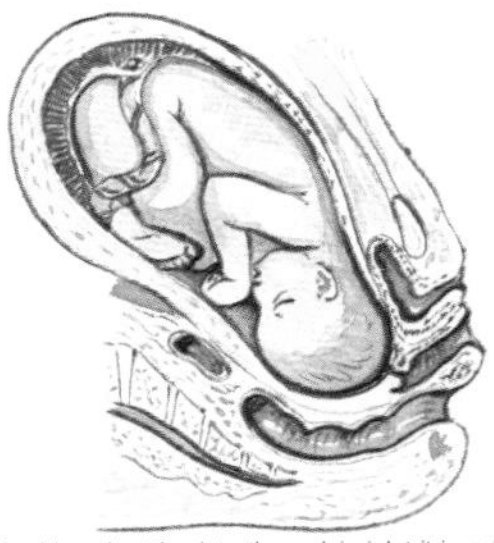

1. When the fetal head settles into the pelvic inlet it is called lightening. It may precede actual birth by three to four weeks in a primapara and a few hours to a week in a multipara.

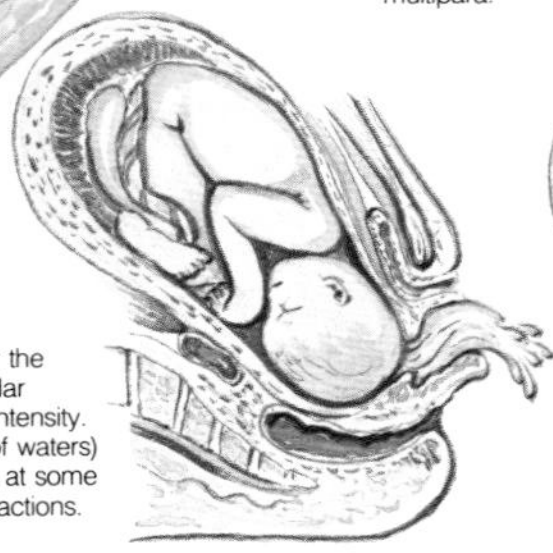

2. In the first stage of labor the uterus undergoes irregular contractions of varying intensity. The amniotic sac (bag of waters) may or may not rupture at some point during these contractions.

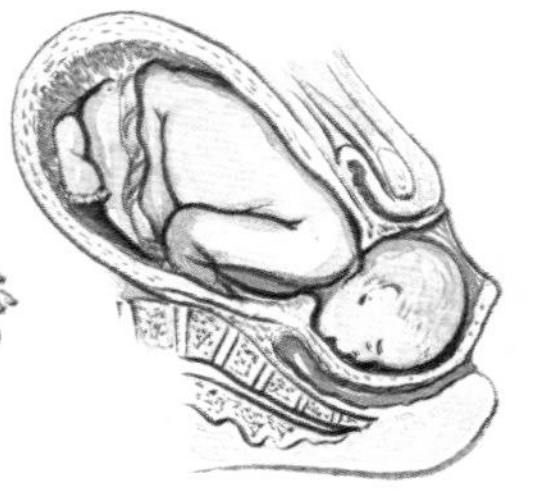

3. Dilation of the cervix.

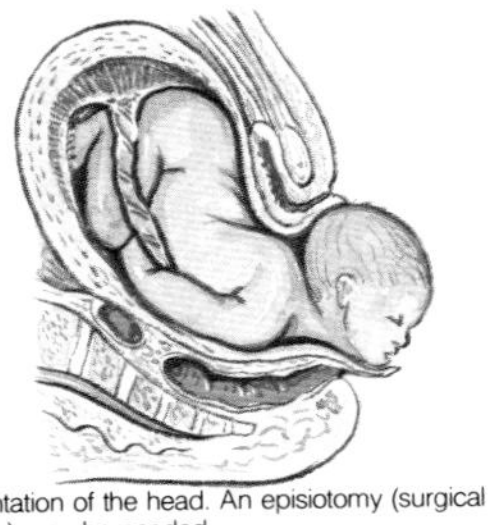

4. Presentation of the head. An episiotomy (surgical incision) may be needed.

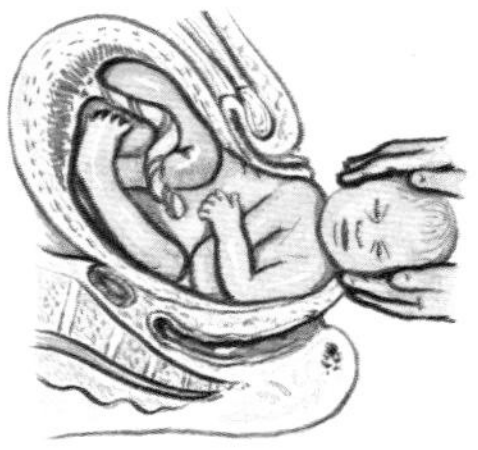

5. Delivery of the head and rotation.

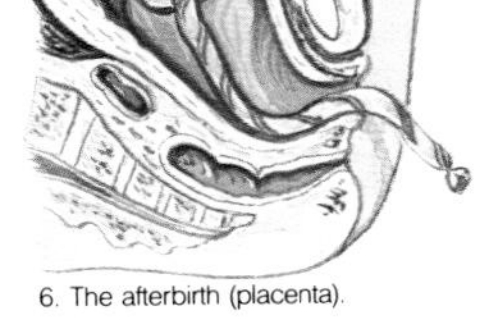

6. The afterbirth (placenta).

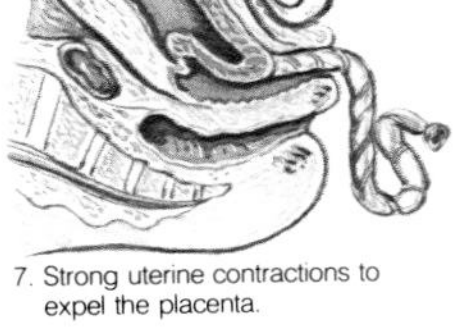

7. Strong uterine contractions to expel the placenta.

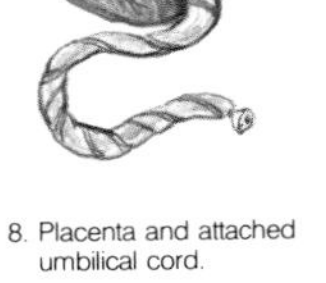

8. Placenta and attached umbilical cord.

CARDIOVASCULAR DISEASE

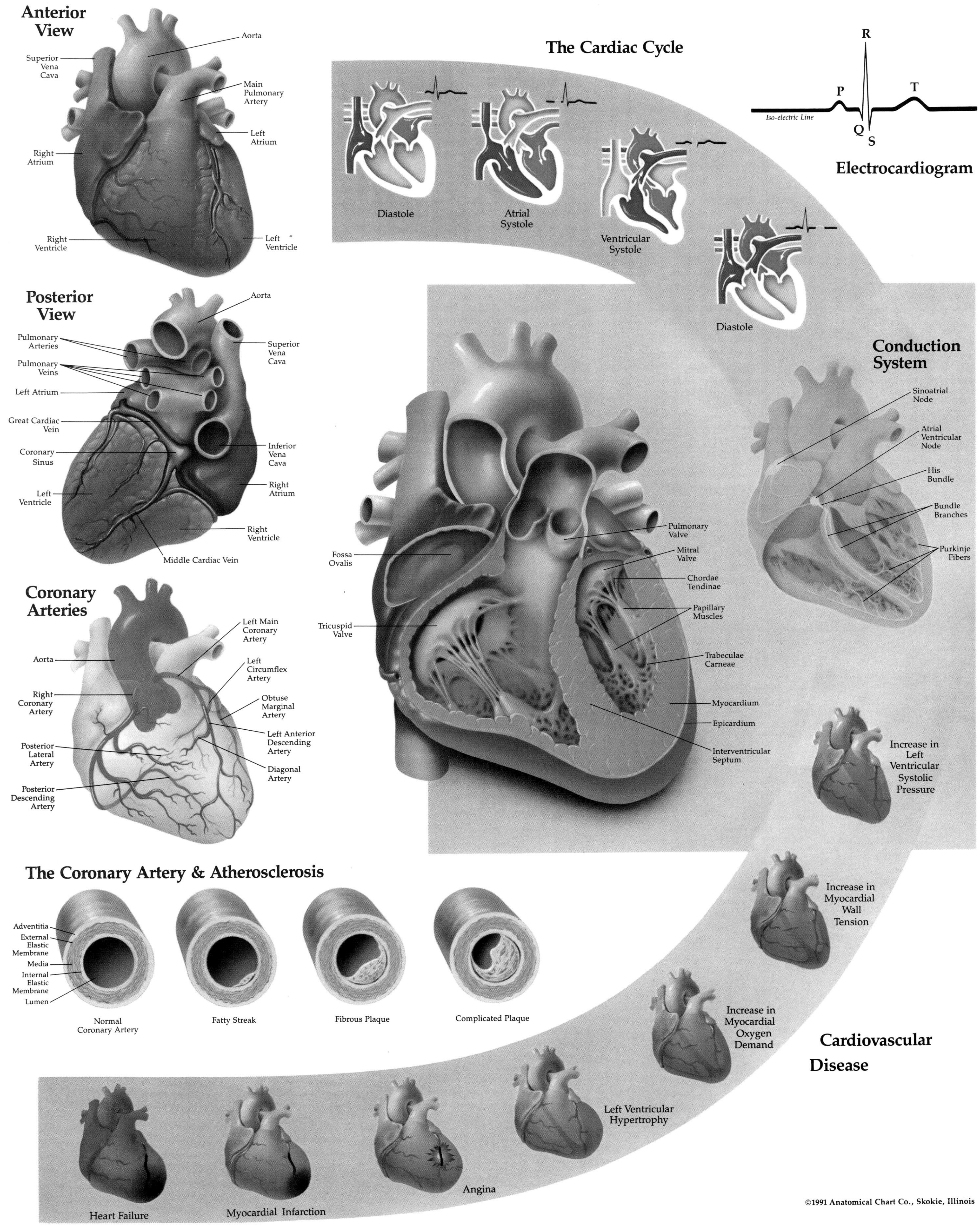
Anterior View
Aorta
Superior Vena Cava
Main Pulmonary Artery
Left Atrium
Right Atrium
Right Ventricle
Left Ventricle
Posterior View
Aorta
Pulmonary Arteries
Superior Vena Cava
Pulmonary Veins
Left Atrium
Great Cardiac Vein
Coronary Sinus
Inferior Vena Cava
Right Atrium
Left Ventricle
Right Ventricle
Middle Cardiac Vein
Coronary Arteries
Left Main Coronary Artery
Aorta
Left Circumflex Artery
Right Coronary Artery
Obtuse Marginal Artery
Left Anterior Descending Artery
Posterior Lateral Artery
Diagonal Artery
Posterior Descending Artery
The Cardiac Cycle
Diastole
Atrial Systole
Ventricular Systole
Diastole
R
P
T
Iso-electric Line
Q
S
Electrocardiogram
Conduction System
Sinoatrial Node
Atrial Ventricular Node
His Bundle
Bundle Branches
Purkinje Fibers
Pulmonary Valve
Mitral Valve
Fossa Ovalis
Chordae Tendinae
Papillary Muscles
Tricuspid Valve
Trabeculae Carneae
Myocardium
Epicardium
Interventricular Septum
Increase in Left Ventricular Systolic Pressure
Increase in Myocardial Wall Tension
Increase in Myocardial Oxygen Demand
Cardiovascular Disease
Left Ventricular Hypertrophy
Angina
Myocardial Infarction
Heart Failure
The Coronary Artery & Atherosclerosis
Adventitia
External Elastic Membrane
Media
Internal Elastic Membrane
Lumen
Normal Coronary Artery
Fatty Streak
Fibrous Plaque
Complicated Plaque

WHIPLASH INJURIES OF THE HEAD AND NECK

Whiplash injury of the head and neck is caused by a sudden exaggerated thrust of the head backward, forward, and sometimes sideways. Abnormal forces are applied to muscles, ligaments, nerves, bones, intervertebral discs, blood vessels, and eyes as the weighty head moves beyond normal physiological limits. There may be no visible bruises or abrasions from this type of injury yet victims report classic symptoms. The symptoms are due to injuries of vertebrae and of soft tissues of the head and neck.

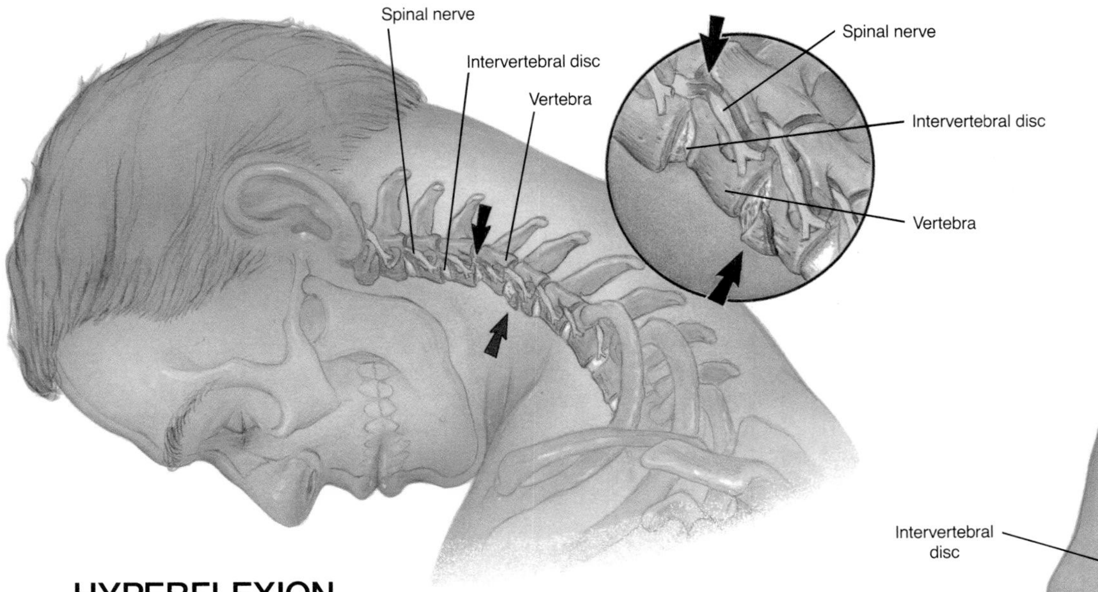

HYPERFLEXION

The head is propelled in a forward and downward motion in hyperflexion. A wedge-shaped deformity of the bone may be created if the anterior portion of the vertebrae are crushed. Intervertebral disks may be damaged. The disks may bulge or herniate, irritating spinal nerves.

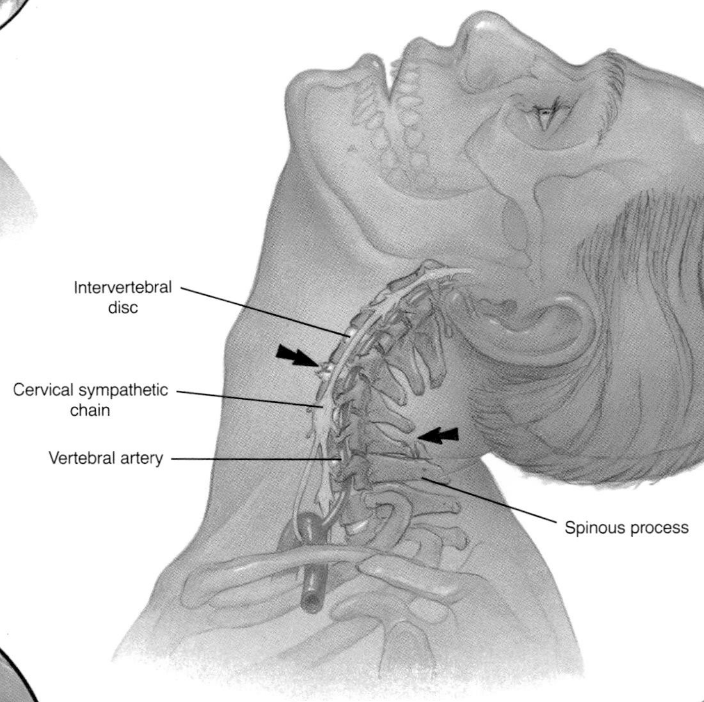

HYPEREXTENSION

The head is forced backward in hyperextension. Pieces of bone may be pulled from cervical vertebrae by a tear of the anterior longitudinal ligament. Spinous processes of the vertebrae may be fractured. Intervertebral disks may be compressed posteriorly and torn anteriorly. Vertebral arteries may be stretched, pinched, or torn, causing an impediment to blood flow to the brain. Nerves of the cervical sympathetic chain may also be injured.

SPINAL LIGAMENTS

Vertebrae are held in place by ligaments. Some ligaments are barely a centimeter long and all are only a few millimeters thick. In a whiplash accident, ligaments may be stretched, partially torn, or completely ruptured.

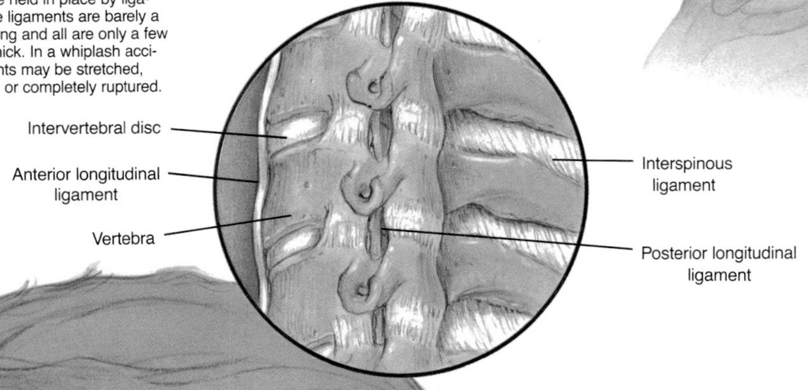

MUSCLE INJURY

Whiplash can cause injuries of neck muscles, ranging from minor strains and microhemorrhages to severe tears. Commonly affected muscles include the sternocleidomastoid muscle, scalene muscles, splenius capitis muscle, and longus colli muscle.

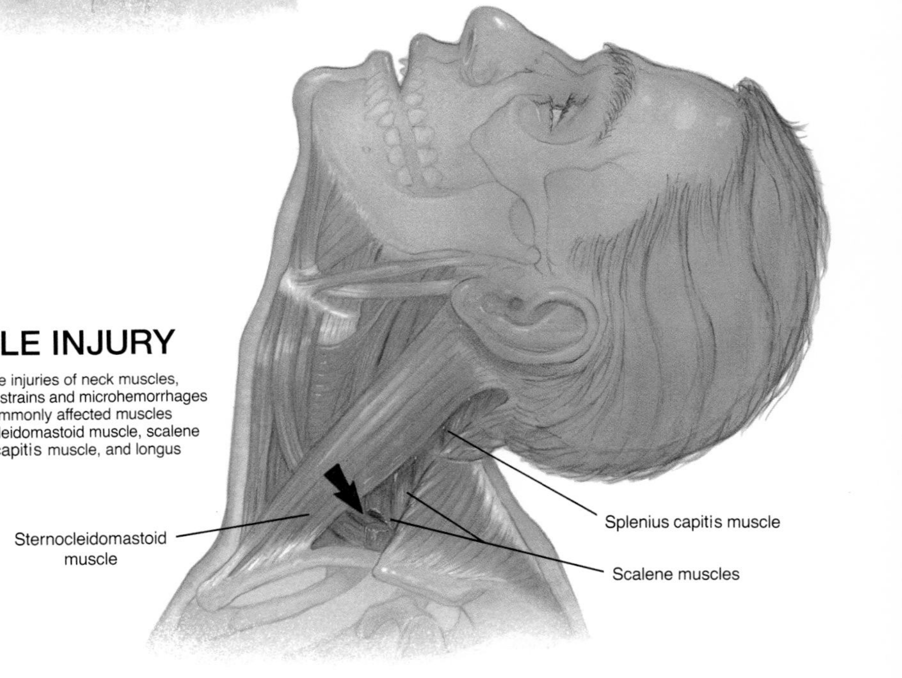

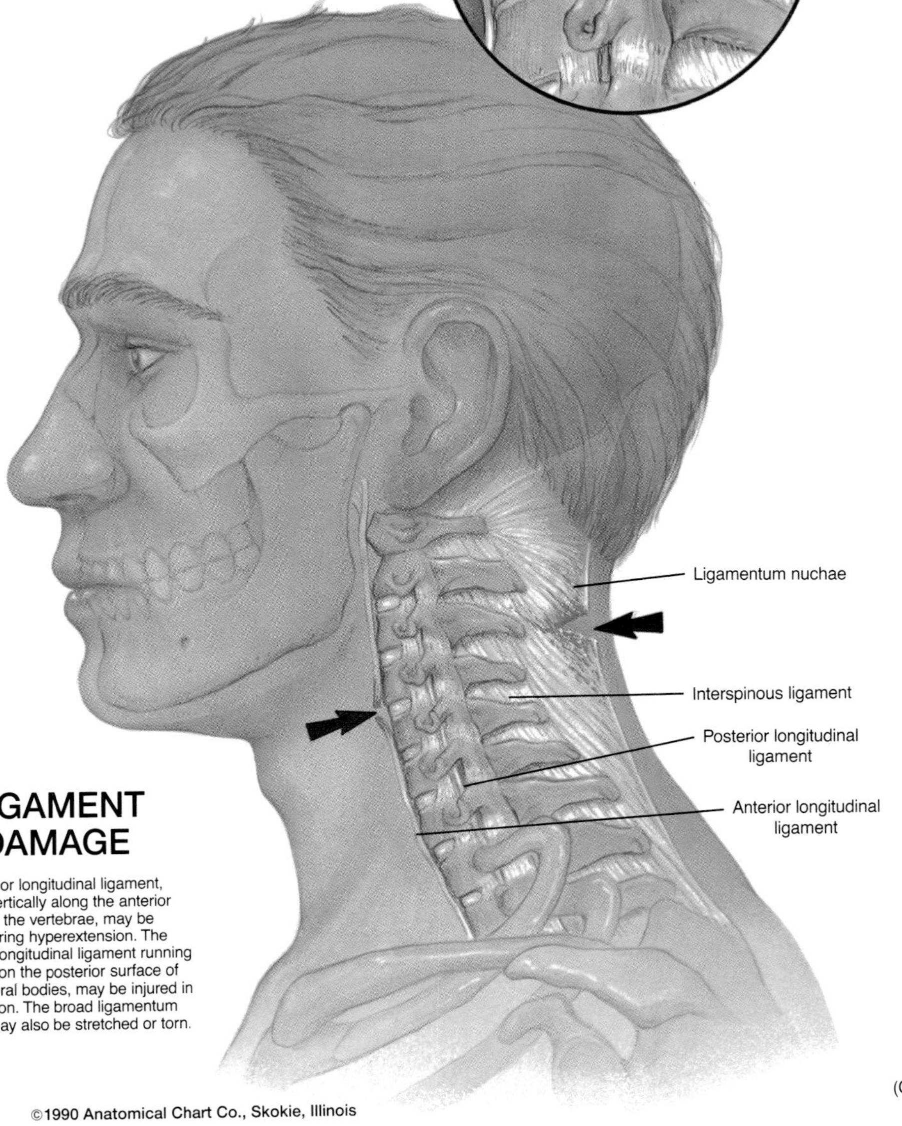

LIGAMENT DAMAGE

The anterior longitudinal ligament, running vertically along the anterior surface of the vertebrae, may be injured during hyperextension. The posterior longitudinal ligament running opposite, on the posterior surface of the vertebral bodies, may be injured in hyperflexion. The broad ligamentum nuchae may also be stretched or torn.

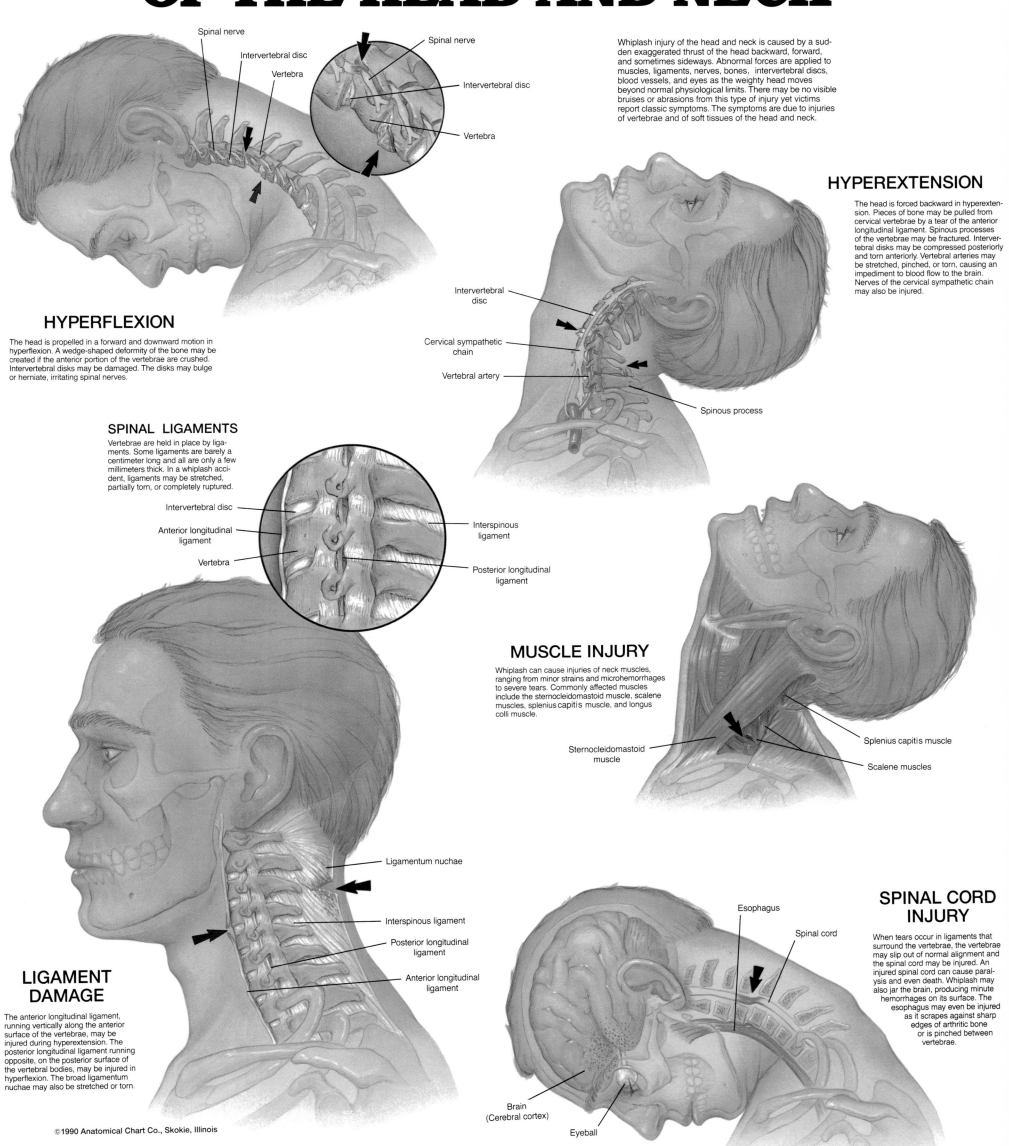

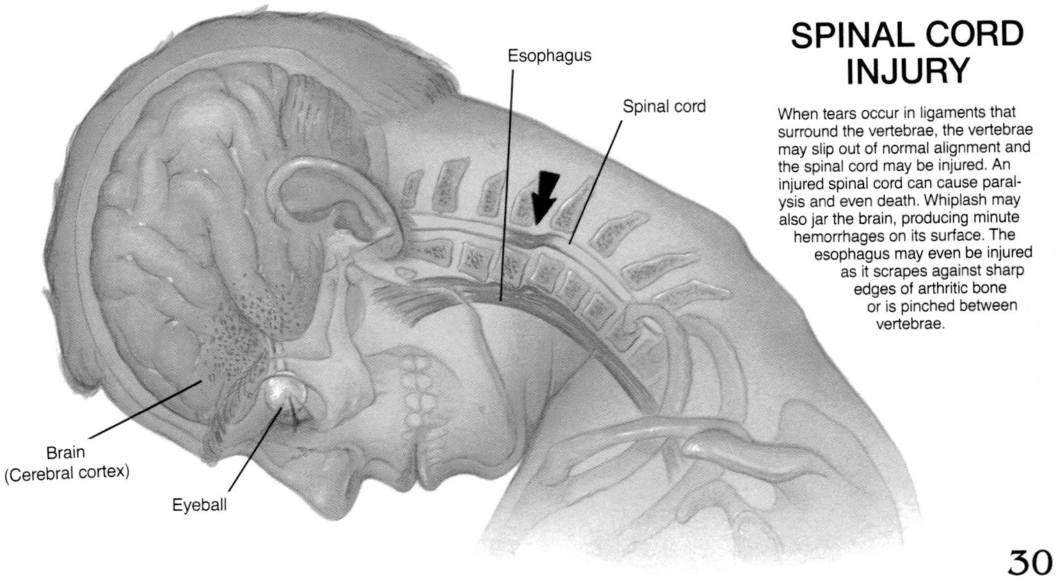

SPINAL CORD INJURY

When tears occur in ligaments that surround the vertebrae, the vertebrae may slip out of normal alignment and the spinal cord may be injured. An injured spinal cord can cause paralysis and even death. Whiplash may also jar the brain, producing minute hemorrhages on its surface. The esophagus may even be injured as it scrapes against sharp edges of arthritic bone or is pinched between vertebrae.

THE HUMAN SPINE - DISORDERS

A TYPICAL CERVICAL VERTEBRA - SUPERIOR ASPECT

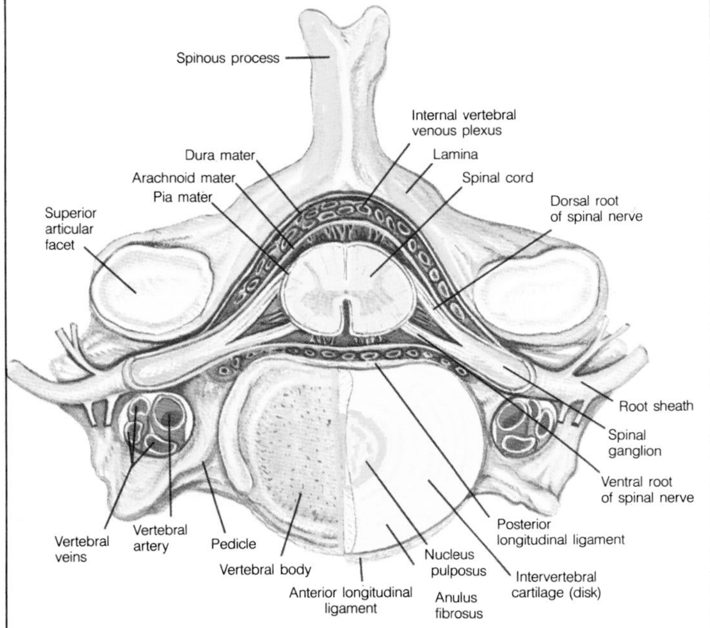

STRUCTURAL FEATURES OF AN INTERVERTEBRAL DISK (SCHEMATIC)

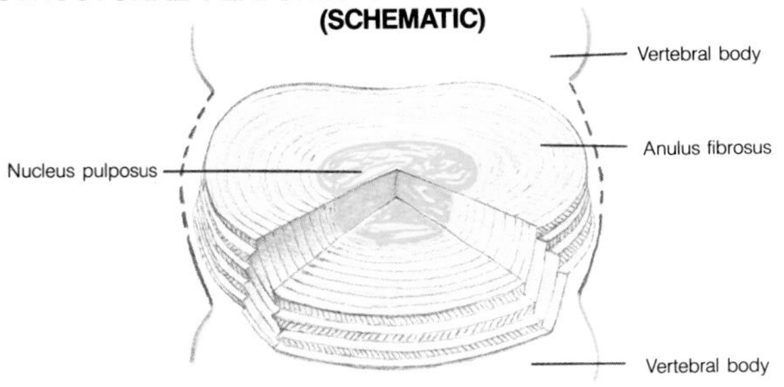

The nucleus pulposus is the central gelatinous part of the intervertebral disk enclosed in several layers of cartilaginous laminae. The nucleus hardens with old age.

Note: alternating obliquity of collagen fibrils

FUNCTION OF INTERVERTEBRAL DISKS

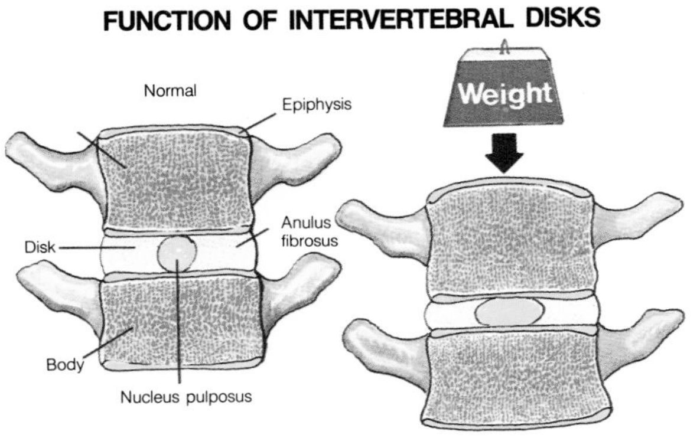

The disk with its contained gelatinous nucleus pulposus serves a cushioning purpose.

SPINAL NERVE IRRITATION DUE TO EXOSTOSIS (BONY OVERGROWTH)

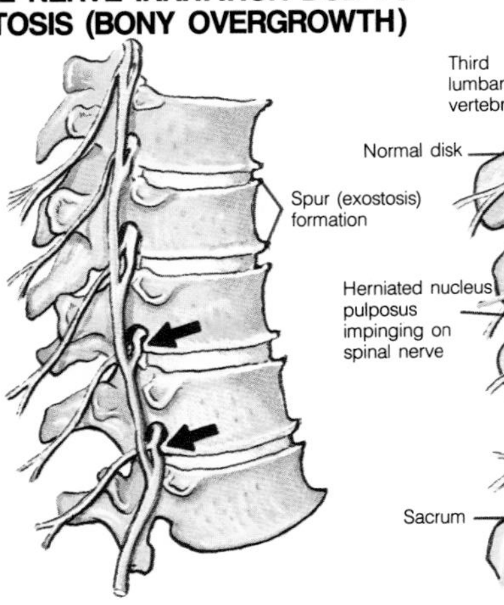

Arrows indicate bone spurs impinging on spinal nerves

A HERNIATED DISK

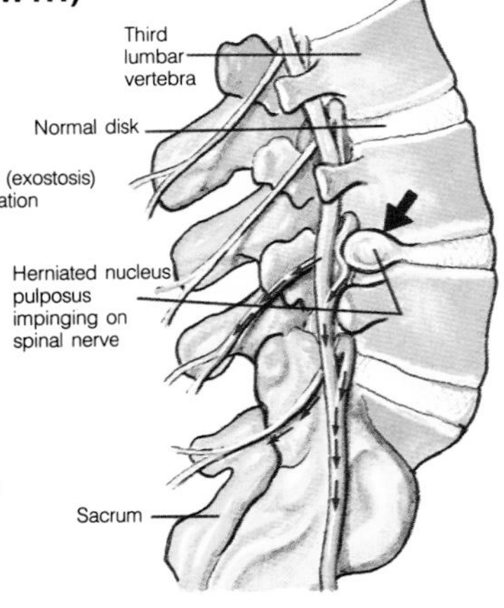

(Arrows indicate perceived pain paths along nerves)

THE SPINAL COLUMN
LATERAL ASPECT

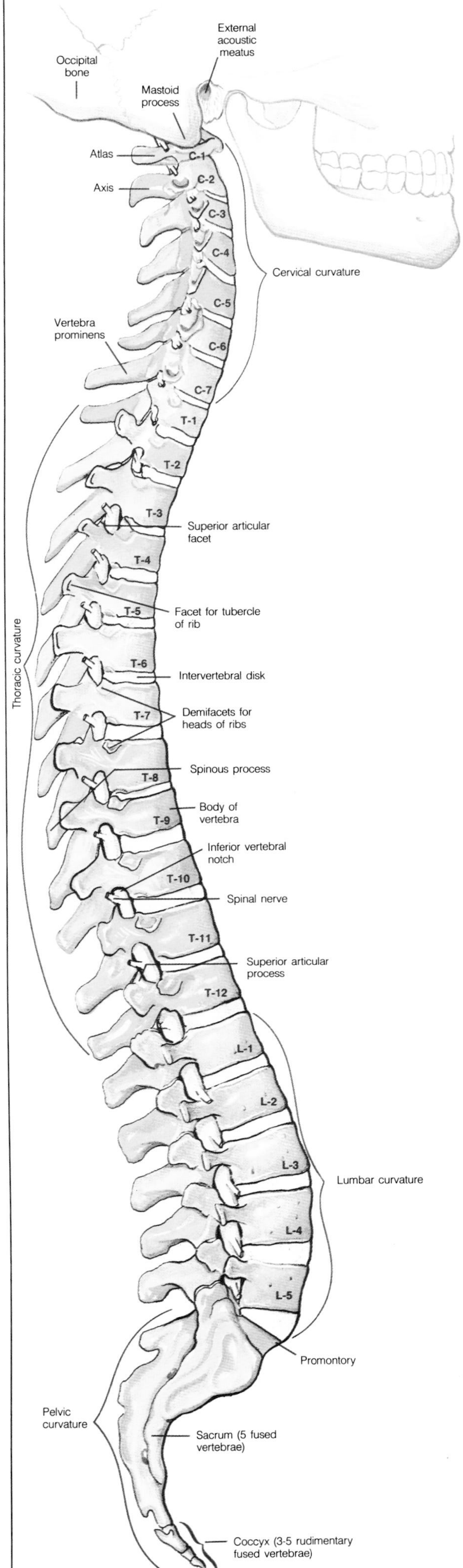

KYPHOSIS

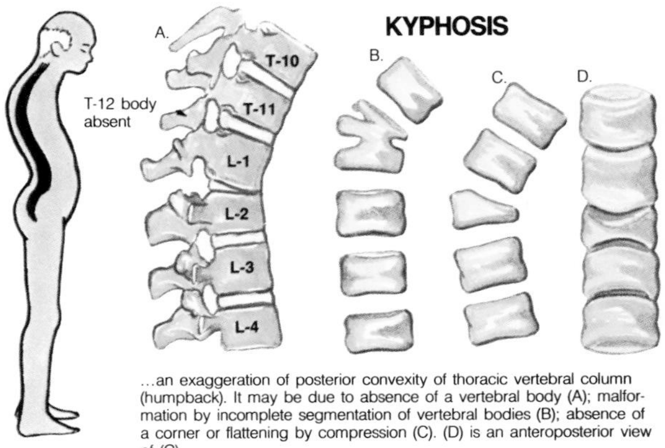

...an exaggeration of posterior convexity of thoracic vertebral column (humpback). It may be due to absence of a vertebral body (A); malformation by incomplete segmentation of vertebral bodies (B); absence of a corner or flattening by compression (C). (D) is an anteroposterior view of (C).

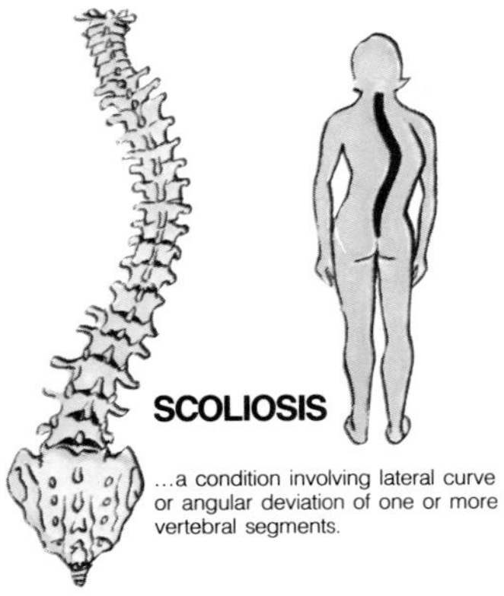

SCOLIOSIS

...a condition involving lateral curve or angular deviation of one or more vertebral segments.

LORDOSIS

...an exaggeration of the posterior concavity of the spine characteristic of the lumbar region. It is also called "swayback" indicating extreme anterior curvature of the lumbar spine.

OSTEOPOROSIS

...a disease of bone due to deficiency of bony matrix.

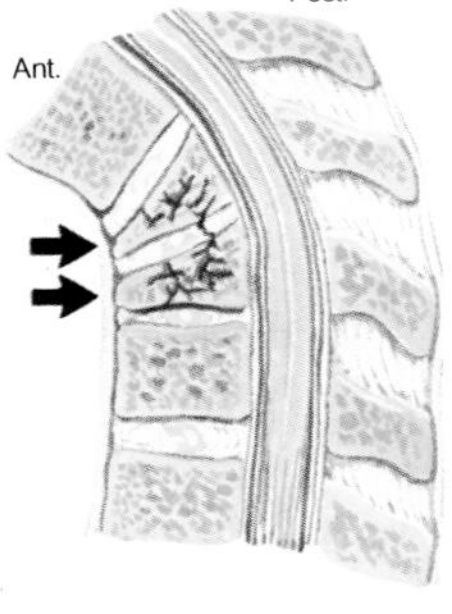

Compression fractures commonly occur at the thoraco-lumbar vertebral junction (arrows) resulting in wedge-shaped vertebrae.

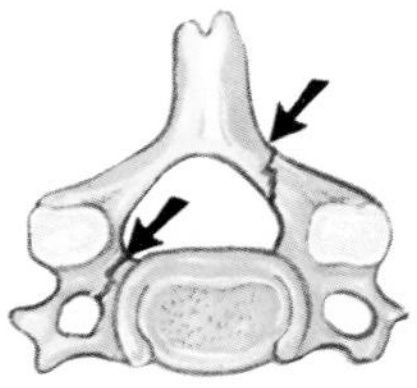

Fractures of laminae or pedicles of the cervical vertebrae are common.

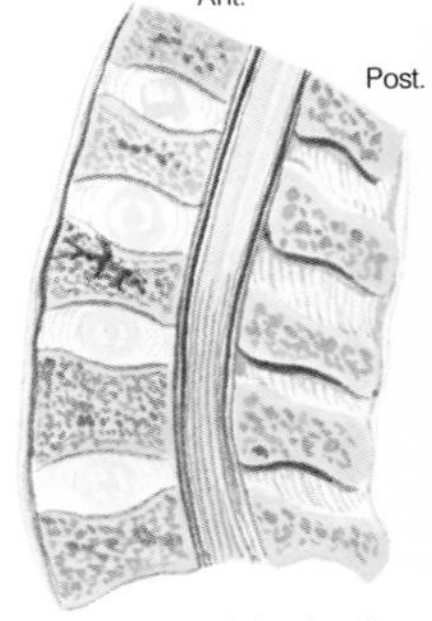

When osteoporosis involves the lumbar region, the vertebral bodies become markedly biconcave and the disks are ballooned.

CAUSES OF PAIN IN THE LOWER BACK OR LEG

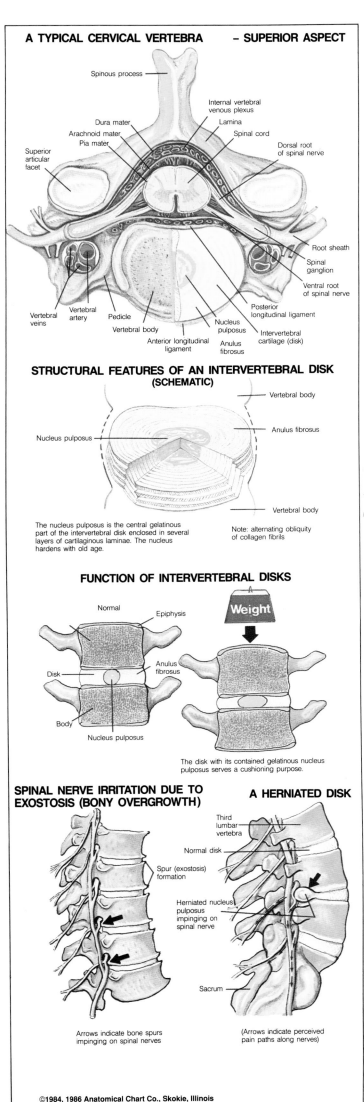

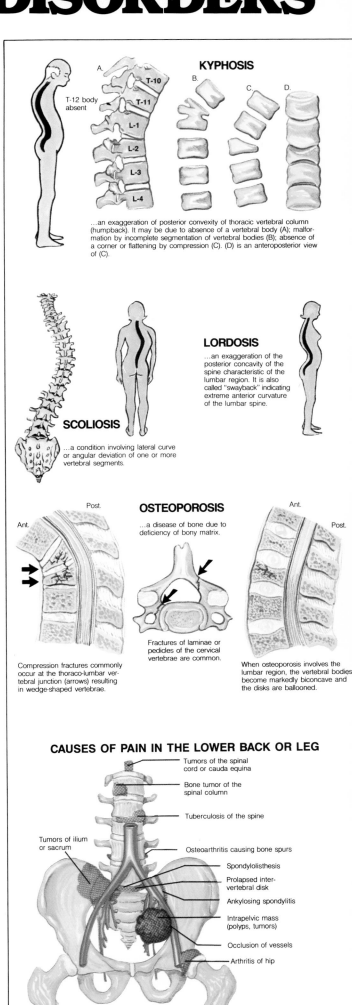

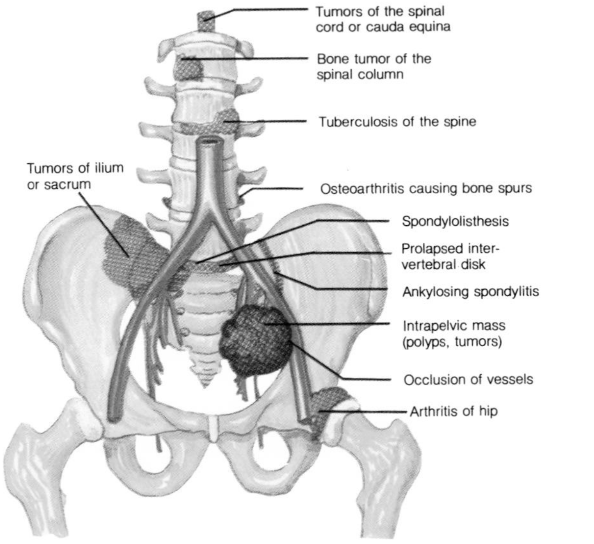

Not all lower back pains are caused by protruding disks or extruded nucleus pulposus. Shown above (diagrammatically) are ten other causes that the examining physician must consider as possibilities in the diagnosis.

DISEASES OF THE DIGESTIVE SYSTEM

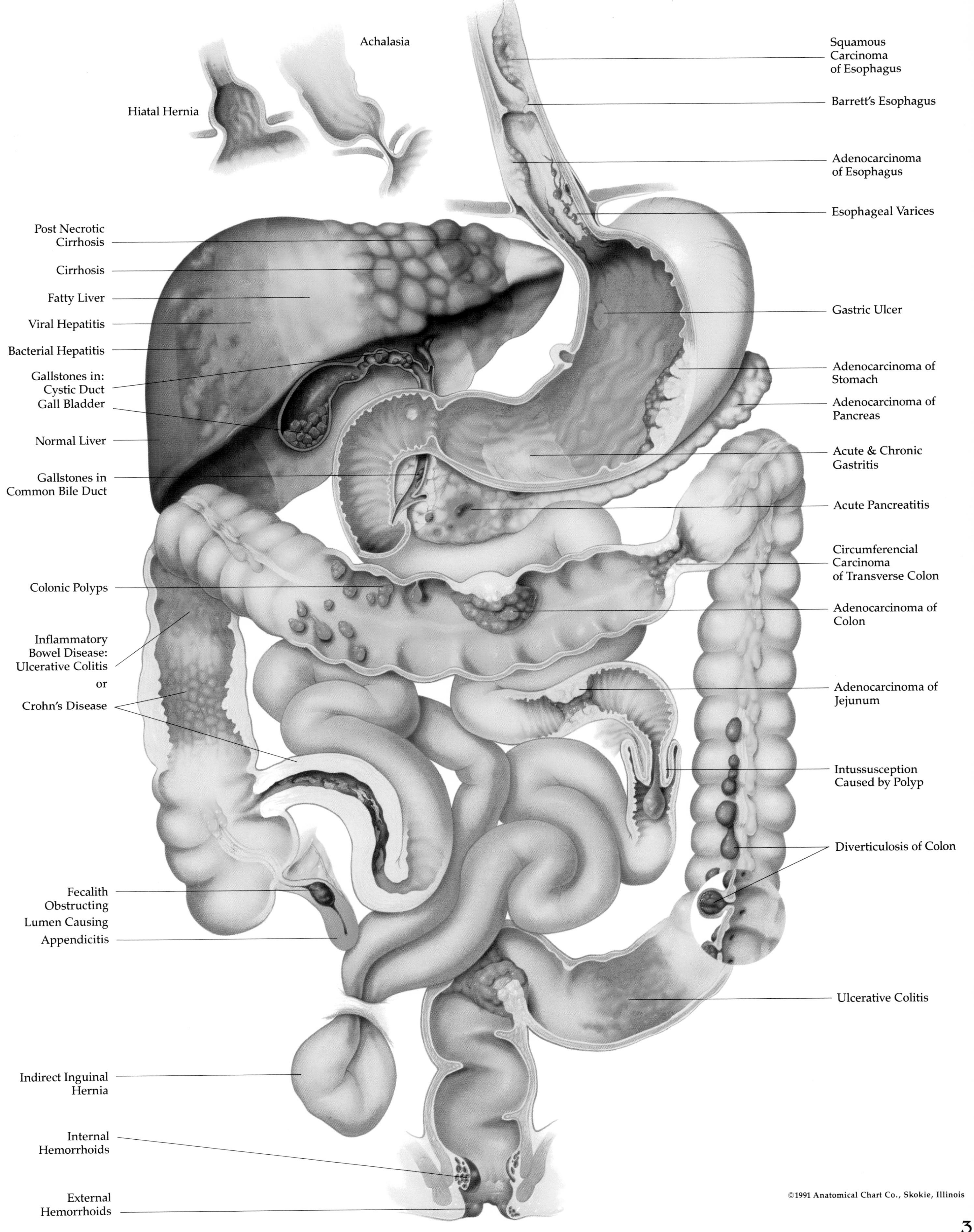
Achalasia
Hiatal Hernia
Squamous Carcinoma of Esophagus
Barrett's Esophagus
Adenocarcinoma of Esophagus
Esophageal Varices
Post Necrotic Cirrhosis
Cirrhosis
Fatty Liver
Viral Hepatitis
Bacterial Hepatitis
Gallstones in: Cystic Duct
Gall Bladder
Normal Liver
Gallstones in Common Bile Duct
Gastric Ulcer
Adenocarcinoma of Stomach
Adenocarcinoma of Pancreas
Acute & Chronic Gastritis
Acute Pancreatitis
Circumferencial Carcinoma of Transverse Colon
Adenocarcinoma of Colon
Colonic Polyps
Inflammatory Bowel Disease: Ulcerative Colitis
or
Crohn's Disease
Adenocarcinoma of Jejunum
Intussusception Caused by Polyp
Diverticulosis of Colon
Fecalith Obstructing Lumen Causing
Appendicitis
Ulcerative Colitis
Indirect Inguinal Hernia
Internal Hemorrhoids
External Hemorrhoids

UNDERSTANDING HYPERTENSION

What is Hypertension?

Hypertension is the result of persistent high arterial blood pressure which may cause damage to the vessels and arteries of the heart, brain, kidneys and eyes. The entire circulatory system is affected since it becomes increasingly more difficult for the blood to travel from the heart to the major organs. Multiple blood pressure readings are taken to establish an average and analyzed by a physician to determine hypertension.

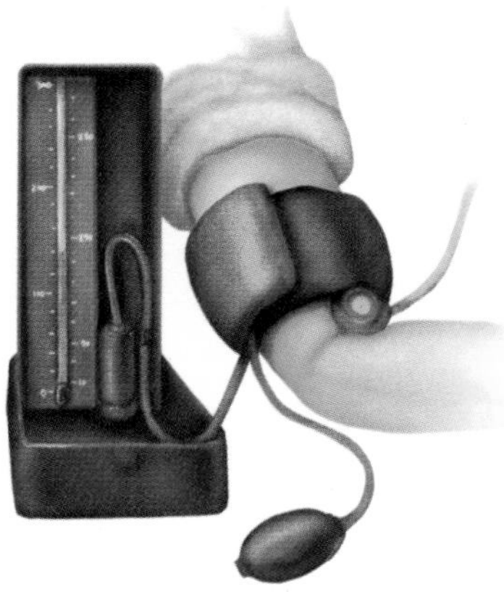

What Causes Hypertension?

Modern life styles tend to increase blood pressure causing hypertension. Some of the known factors include a high salt intake, excessive alcohol consumption and obesity. Genetic factors may also influence this disease. Primary hypertension is the most common type and it generally is improved by a healthier life style; and medication when needed. Secondary hypertension is the result of a disorder or abnormality of the kidney, adrenal gland or other vital organ. This less common type of hypertension is often treated surgically. Hypertension may also occur during pregnancy and requires special attention.

What is Blood Pressure?

Blood pressure is a measure of the pressure of the blood against the walls of the arteries. It is dependent upon the action of the heart, the elasticity of the artery walls and the volume and thickness of the blood. The blood pressure readings are a ratio of the maximum or systolic pressure (as the heart pushes the blood out to the body) written over the minimum or diastolic pressure (as the heart begins to fill with blood).

$$\frac{\text{Systolic pressure}}{\text{Diastolic pressure}} \quad \text{or} \quad \frac{120}{80}$$

Symptoms of Hypertension

You may have:

NO SYMPTOMS!
(No noticeable symptoms may be felt even with high blood pressure)

or you may have:

- Headaches
- Blurring of vision
- Chest pain
- Frequent urination at night

Effects in Blood Vessels

Increases in arterial blood pressure can change and damage the inside artery wall. The wall may become thicker while the space which transports the blood becomes smaller (vascular hypertrophy).

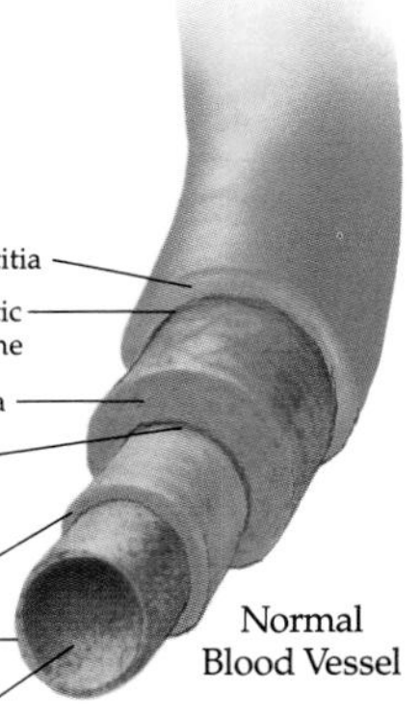

Normal Blood Vessel

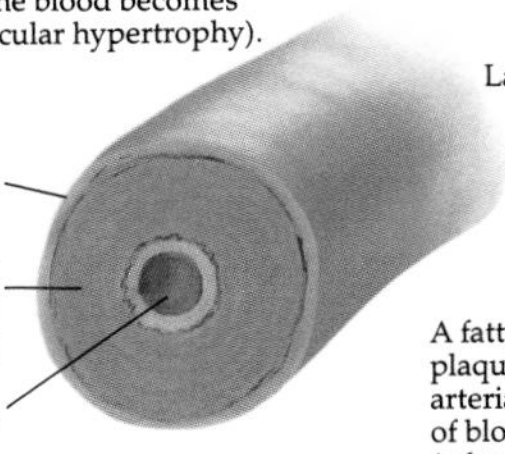

Vascular Hypertrophy

A fatty build up, also called plaque, develops in the damaged arterial wall, clogging the flow of blood throughout the artery (atherosclerosis). Blood clots may form more easily and become dangerous if dislodged.

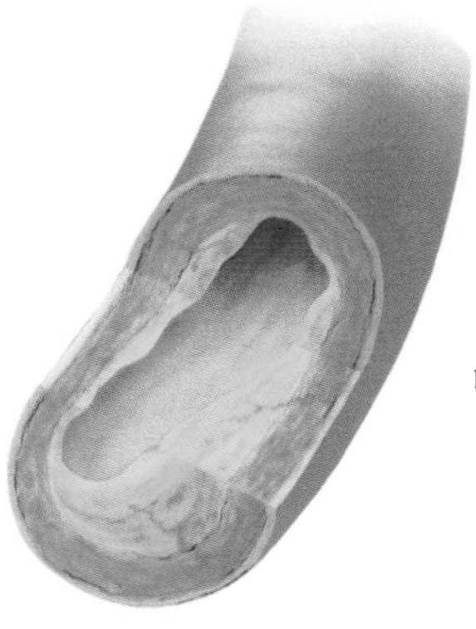

Atherosclerosis

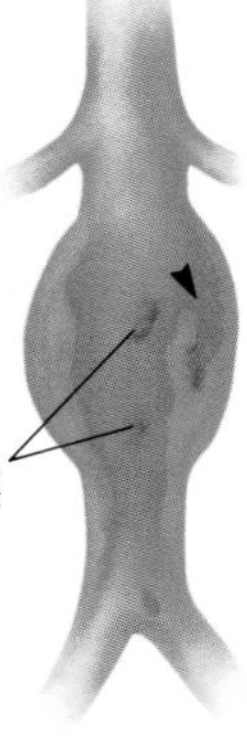

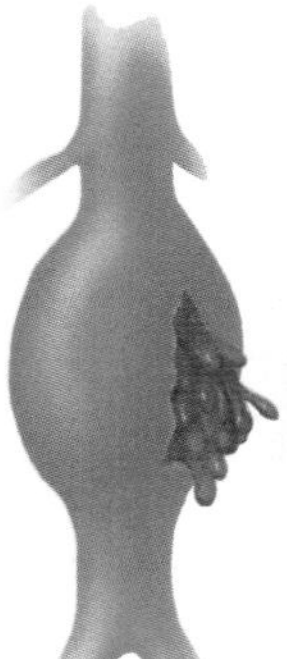

Under increasing blood pressure, a weakening of the artery wall may balloon out (aneurysm) and break, causing blood loss, tissue damage and even death.

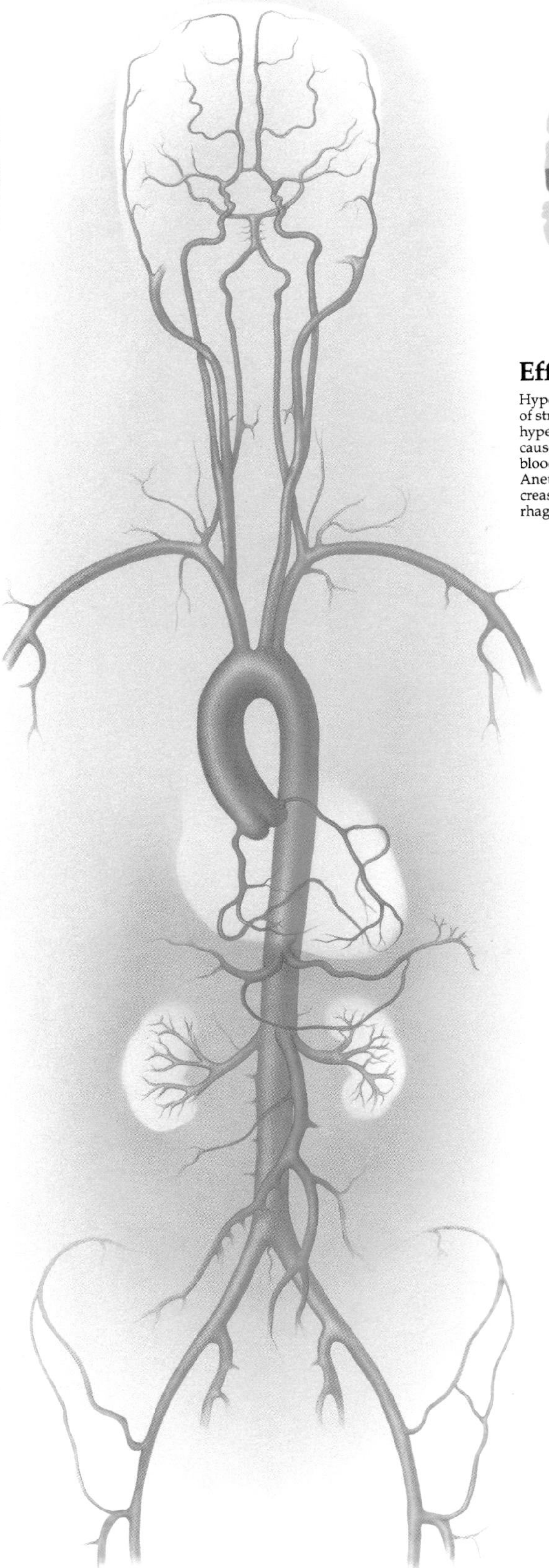

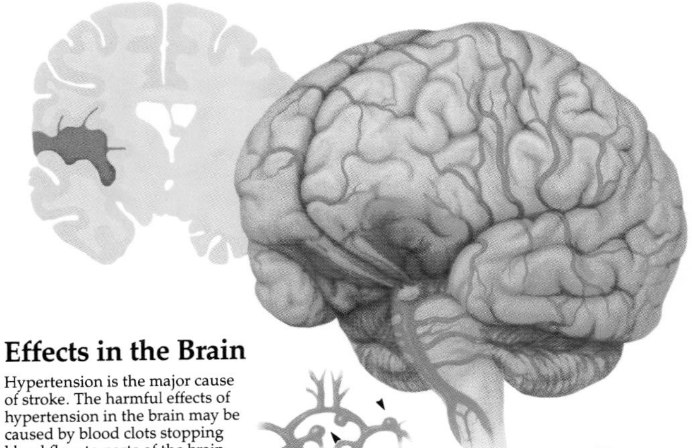

Effects in the Brain

Hypertension is the major cause of stroke. The harmful effects of hypertension in the brain may be caused by blood clots stopping blood flow to parts of the brain. Aneurysms may burst under increasing pressure causing hemorrhage and damage to brain tissue.

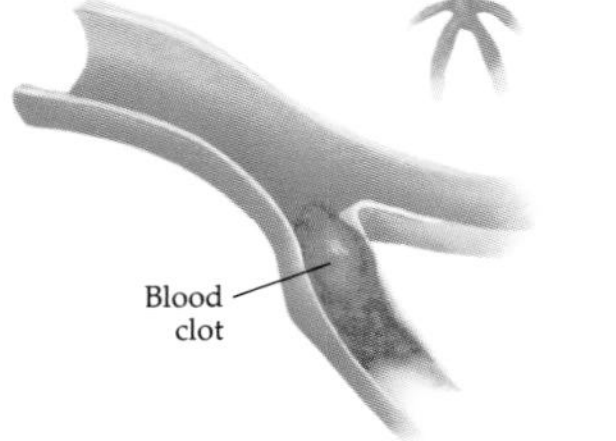

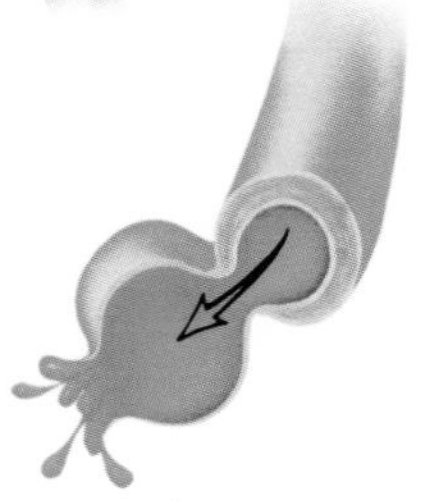

Aneurysm

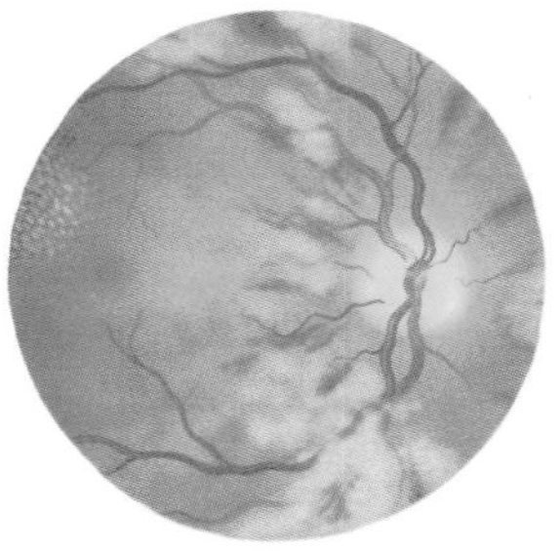

Effects in the Eye

A thorough eye examination by a physician may lead to the diagnosis of hypertension. This can be determined by the vascular changes in the back of the eye (retina).

Effects in the Kidneys

The kidneys are easily damaged by hypertension. In addition, many kidney diseases cause hypertension. Increased blood pressure disrupts the kidneys' ability to regulate salt and water balance in the body which can make hypertension worse.

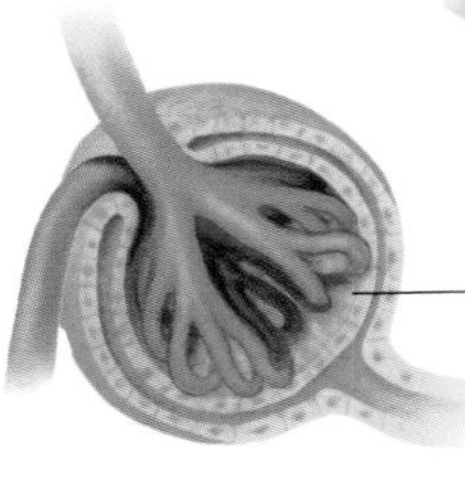

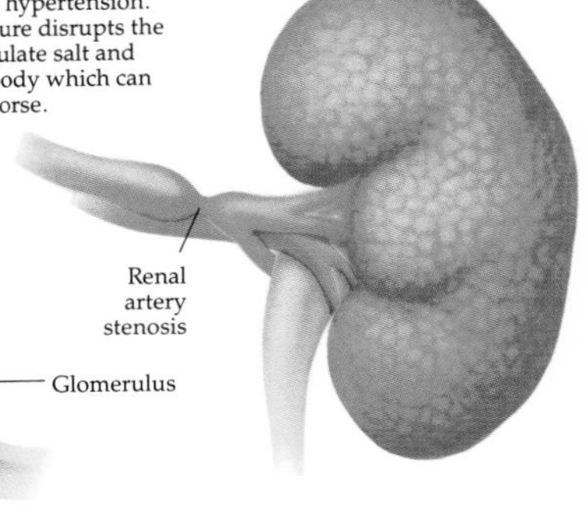

Blood Flow in the Heart

The right side of the heart receives blood from the body and delivers this unoxygenated blood to the lungs. The left side of the heart receives oxygen rich blood from the lungs and pumps it through the arteries to all organs and tissues in the body.

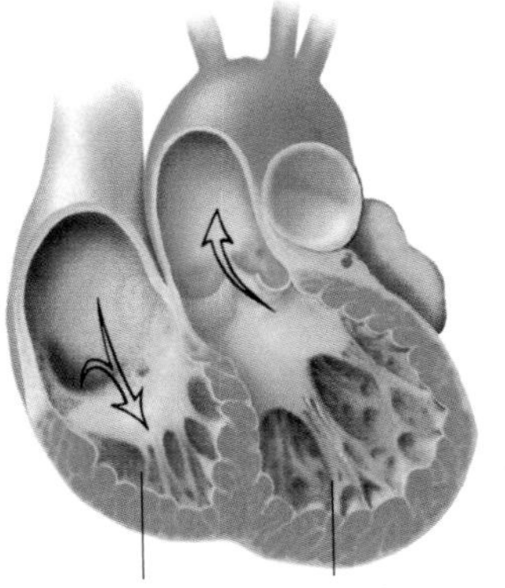

Normal Heart

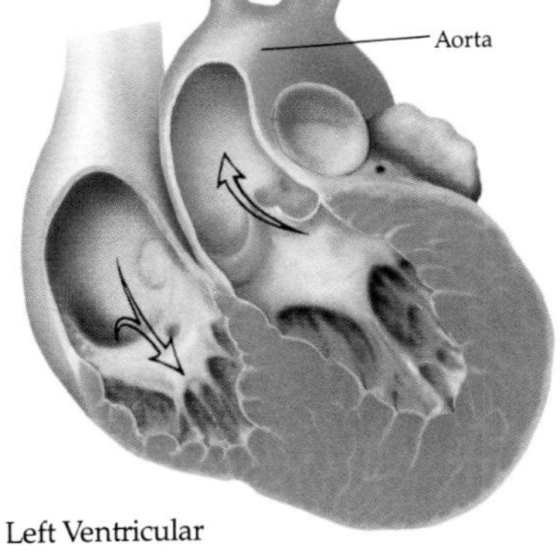

Left Ventricular Hypertrophy

Effects in the Heart

Hypertension can cause serious health problems to this vital organ. Increased resistance in the arteries, due to stiffness and narrowing of the vessels causes the left heart to work harder pumping against a higher pressure (vascular hypertrophy). The left ventricle may become enlarged and unable to respond to this pressure increase. In addition, the heart muscle may suffer from decreased blood flow due to atherosclerosis of the small arteries of the heart.

Healthy Life Style Changes

Decrease your blood pressure by:

- Reducing body weight
- Restricting dietary salt
- Increasing fiber and decreasing fat in your diet
- Not smoking
- Avoiding excess alcohol
- Exercising regularly
- Developing relaxation techniques

It is very important to follow your physician's instructions and to take any medications as prescribed.

UNDERSTANDING STROKE

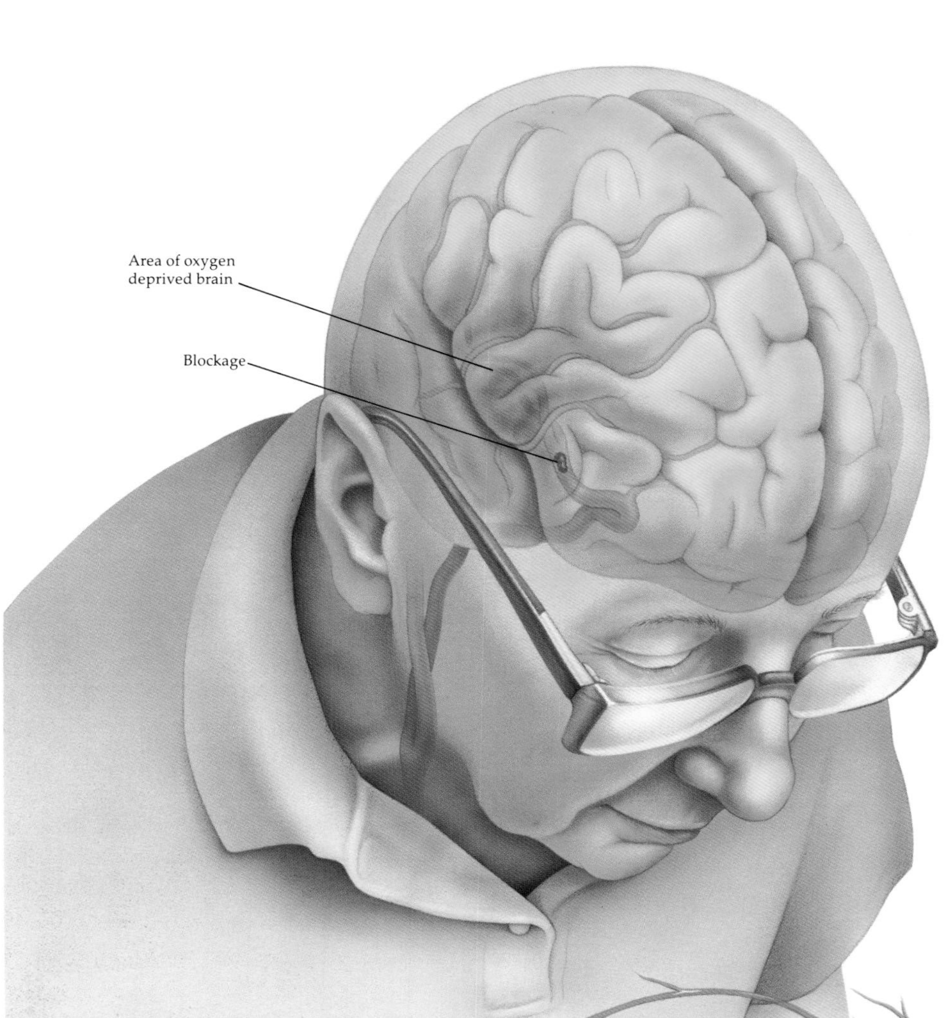

What is Stroke?

The term stroke refers to the sudden death of brain tissue caused by a lack of oxygen resulting from an interrupted blood supply. An **infarct** is the area of the brain which "died" because of this lack of oxygen. There are two ways that brain tissue death can occur. **Ischemic stroke** is a blockage or reduction of blood flow in an artery that feeds that area of the brain. It is the most common cause of an infarct. **Hemorrhagic stroke** results from bleeding within and around the brain causing compression and tissue injury.

Events Leading to Stroke

Stroke victims often have small strokes or "warning signs," before a large permanent attack.

Transient Ischemic Attacks (TIAs) are brief attacks that last anywhere from a few minutes to 24 hours. The symptoms resolve completely and the person returns to normal. It is possible to have several TIAs before a large attack.

Reversible Ischemic Neurological Deficit (RIND) is an attack that lasts longer than 24 hours with recovery usually within 3 weeks. No tissue death occurs during a RIND, but the risk of a complete infarction is greatly increased after one.

Complete Infarction (CI) is an attack that leaves permanent tissue death and results in serious neurological deficits. Recovery is usually not total and lasts longer than 3 weeks.

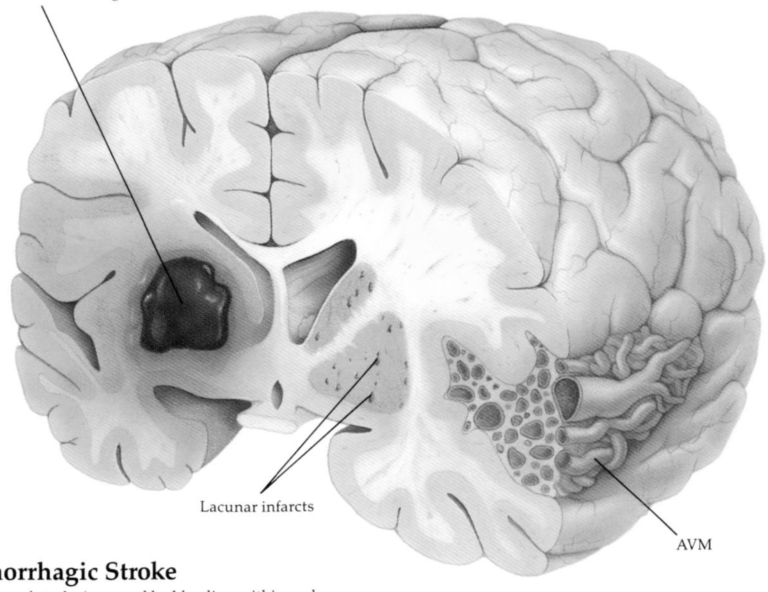

Hemorrhagic Stroke

This type of stroke is caused by bleeding within and around the brain. Bleeding that fills the spaces between the brain and the skull is called a subarachnoid hemorrhage. It is caused by ruptured aneurysms, arteriovenous malformations, and head trauma. Bleeding within the brain tissue itself is known as intracerebral hemorrhage and is primarily caused by hypertension.

An **arteriovenous malformation** (AVM) is an abnormality of brain blood vessels where arteries lead directly into veins without first going through a capillary bed. The pressure of the blood coming through the arteries is too high for the veins causing them to dilate to transport the higher volume of blood. This dilation can cause them to rupture or bleed.

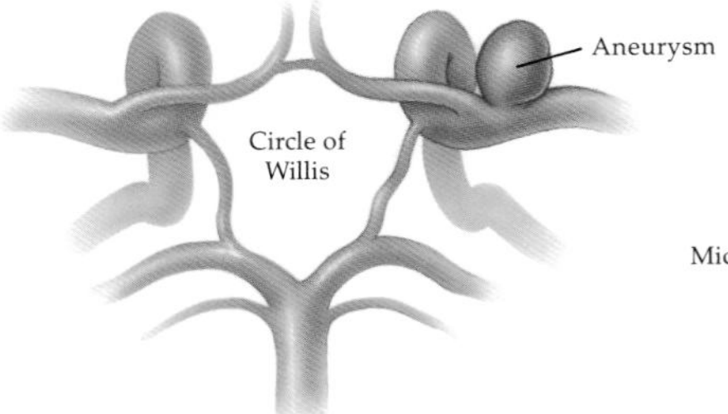

An **aneurysm** is a weakening of the arterial wall causing it to stretch and balloon. It usually occurs where the artery branches.

Hypertension is an elevation of blood pressure which may cause tiny arterioles to burst in the brain. Blood released inside brain tissue puts pressure on adjacent arterioles causing them to burst and lead to more bleeding. Hypertension may also cause lacunar infarcts. These are miniature infarcts similar to complete strokes, but on a much smaller scale. They occur within the nuclei and spinal tracts of the brain and resemble little lakes or pits.

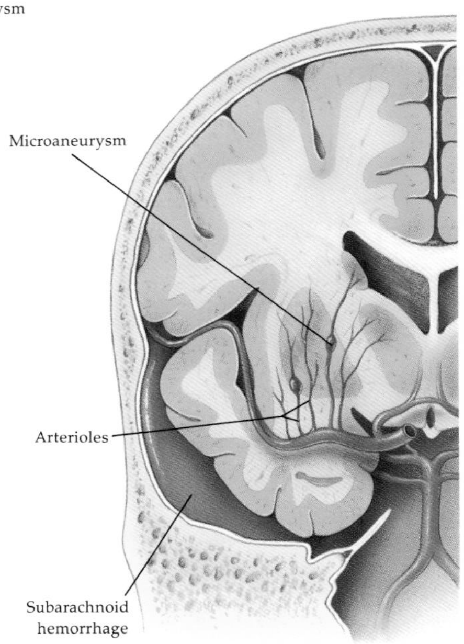

Ischemic Stroke

This type of stroke results from a blockage or reduction of blood flow to an area of the brain. This blockage may result from atherosclerosis and blood clot formation.

Atherosclerosis, or hardening of the arteries, is the deposit of cholesterol and plaques within the wall of arteries. These deposits may become large enough to narrow the lumen and reduce the flow of blood while also causing the artery to lose its ability to stretch.

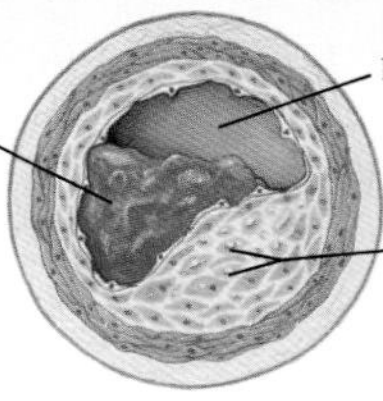

A **thrombus**, or blood clot, forms on the roughened surface of atherosclerotic plaques that develop in the wall of the artery. The thrombus can enlarge and eventually block the lumen of the artery.

Part of a thrombus may break off and become an **embolus.** An embolus travels through the bloodstream and may block smaller arteries.

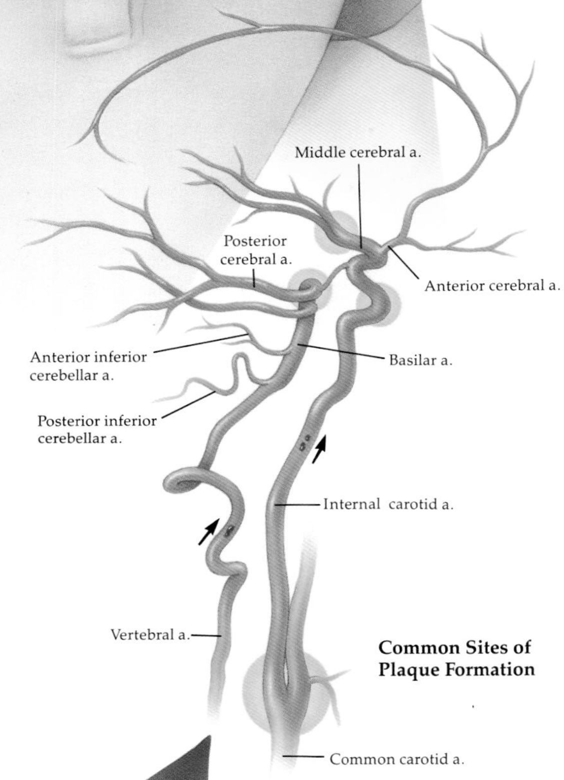

Common Sites of Plaque Formation

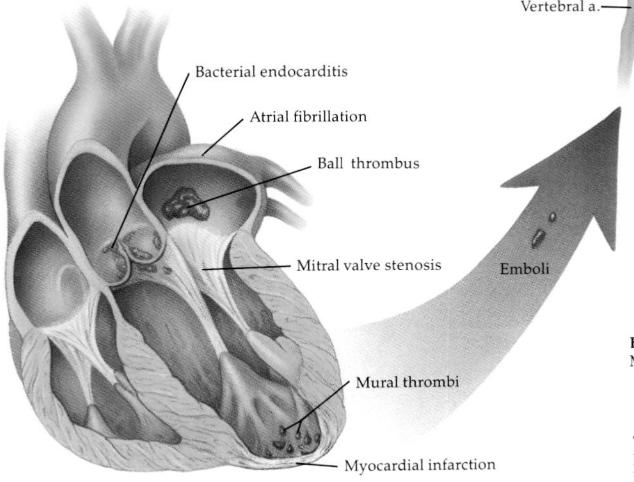

Emboli commonly come from the heart, where different diseases can cause thrombi formation.

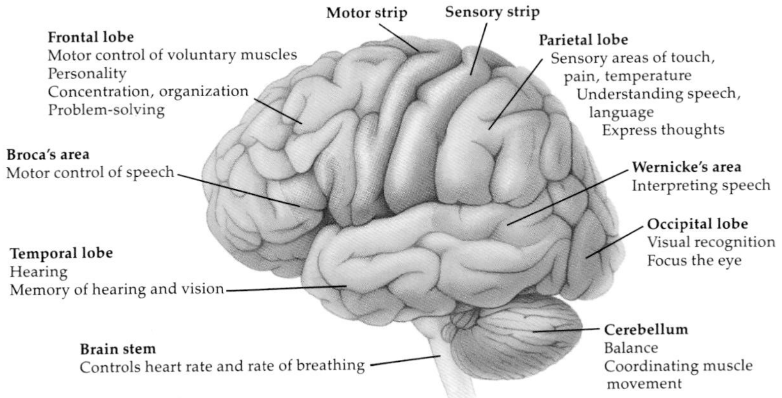

Normal Functional Areas of Brain

The brain has two sides: a right hemisphere that controls the left side of the body and a left hemisphere that controls the right side of the body. Each hemisphere has four lobes and a cerebellum that control our daily functions. Depending on what part of the brain has been affected, stroke victims experience a variety of neurological deficits. Rehabilitation is crucial to the stroke patient. Physical therapists and speech therapists will help patients "relearn" their lost functions and devise ways to cope with those they cannot regain.

What Puts You At Risk

Hypertension
Heart disease
Atherosclerosis
Previous TIAs
High cholesterol
High alcohol consumption
Obesity
Diabetes
Bruit noise in carotid artery
Cigarette smoking
Oral contraceptives
Family history

Common Neurological Deficits After Stroke

Left-sided stroke	Right-sided stroke
Right-sided paralysis	Left-sided paralysis
Speech/language deficits	Spacial/perceptual deficits
Slow, cautious behavior	Quick, impulsive behavior
Hemianopia of right visual field	Hemianopia of left visual field
Memory loss in language	Memory loss in performance
Right-sided dysarthria	Left-sided dysarthria
Aphasia	
Apraxia	

Related Terms

Paralysis	Loss of muscle function and sensation
Hemiparesis	Weakness of muscles on one side of body
Hemianopia	Loss of sight in half of visual field
Aphasia	Unable to understand or produce language
Apraxia	Unable to control muscles, movement is uncoordinated and jerky
Dysarthria	Slurring of speech and "mouth droop" on one side of face due to muscle weakness

Medical Illustrations by Tonya Hines.

UNDERSTANDING HIV & AIDS

What is AIDS?

Acquired Immunodeficiency Syndrome or AIDS is a devastating condition which develops after the immune system has been severely weakened. The immune system collapses and can no longer defend the body from certain infections. As AIDS progresses, the body becomes overwhelmed with life threatening illnesses, diseases and cancers.

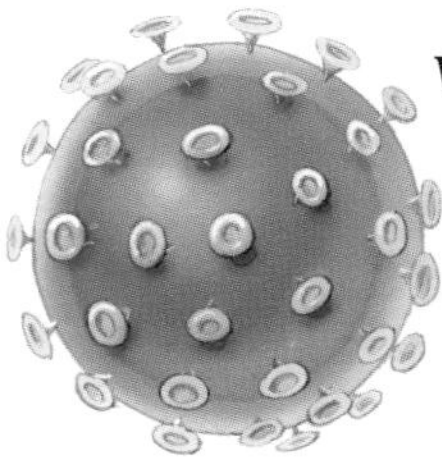

What is HIV?

HIV is the virus that causes AIDS. It stands for Human Immunodeficiency Virus and is also referred to as the AIDS virus. HIV destroys immune cells called T-cells, which are vital to the body for protection against infection, disease and cancer. AIDS results when HIV has severely reduced the number of T-cells.

HIV (AIDS virus)

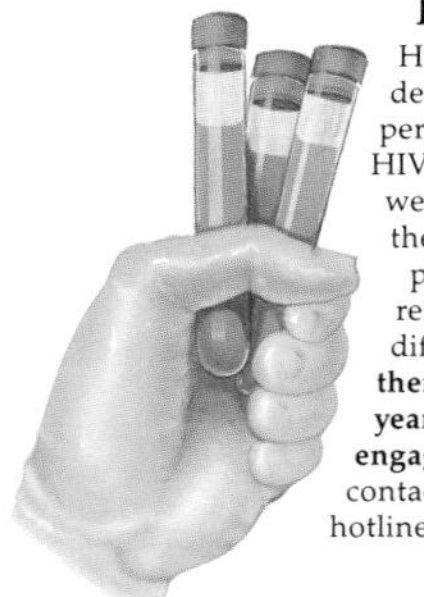

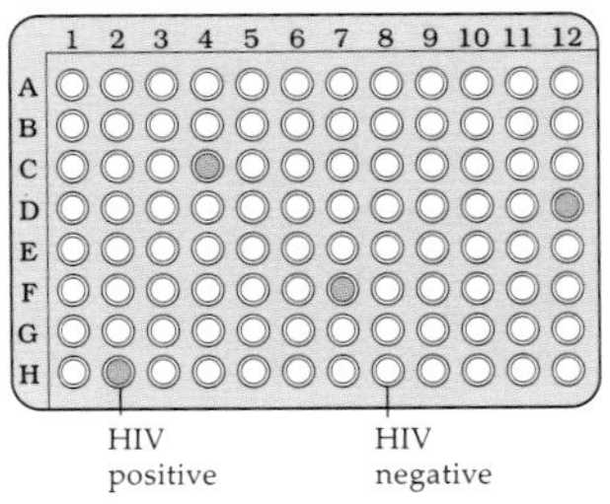

Who Develops AIDS?

Everyone that is infected with HIV has the possibility of developing AIDS, and anyone can become infected with HIV. **AIDS is not a homosexual disease, it is a human disease that can infect anyone.** AIDS does not see race, religion, color or sexual orientation, only the opportunity to infect bodies through the contact of infected blood, semen or vaginal secretions. Even the unborn can be infected.

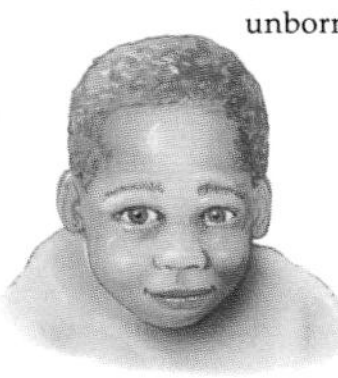

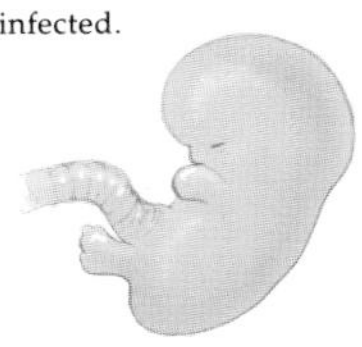

How is HIV Transmitted?

HIV lives in white blood cells, called T-cells, which are found in blood, semen and vaginal secretions including menstrual blood. HIV can be transmitted when there is contact with infected fluids. The most common ways people become infected are through sexual activities, injecting drug use, pregnancy and by receiving infected blood or clotting factors.

Sexual Activities

Sexual activity is the most common way HIV is spread, especially without the use of a latex condom. **Contact with infected blood, semen, vaginal secretions or menstrual blood, can lead to the transmission of HIV.** The Virus can also be transmitted through oral sex.

This represents vaginal or rectal tissues

Sperm   HIV

HIV Antibody Test

HIV antibody testing determines HIV infection by detecting HIV antibodies in the blood of an infected person. HIV antibodies are produced in response to HIV infection and can be detected in the blood 4 to 12 weeks after infection. An HIV negative result means there were no antibodies found in the blood; HIV positive means antibodies are present. A positive result is always confirmed by another test using a different method. **Other than the AIDS antibody test, there are often no signs or symptoms of infection for years, which is why it is important to be tested if engaged in high risk behavior.** For testing locations contact your physician or call the National AIDS hotline (1•800•342•2437)

Preliminary HIV antibody test results

```
    1  2  3  4  5  6  7  8  9 10 11 12
A
B
C
D
E
F
G
H
```
HIV positive   HIV negative

Each circle represents one person

HIV Positive

A positive test for HIV means you have been infected with HIV for life and are able to infect others. Inform those whom you have engaged in sexual activities with or shared needles. Stay healthy, because even a common cold or flu can help the virus weaken the immune system. **Not all people with HIV develop AIDS; however, it is likely to develop.** Take added precautions during sexual activities and drug use to prevent spreading HIV.

Symptoms (of HIV infection)

Memory loss
Disorientation
Inability to think clearly

White patches on tongue

Swollen lymph nodes in neck, armpits and groin

Heavy night sweats

Loss of appetite

Severe weight loss

Diarrhea

Fatigue & muscle weakness

Vein
HIV
White blood cell
HIV antibodies

AIDS (related illnesses)

Cryptococcal menengitis Inflammation of tissue around the brain and central nervous system

Toxoplasmosis encephalitis It is the most common central nervous system OI

Cytomegalovirus retinitis (CMV) Leads to loss of vision

Herpes simplex virus (HSV) Sores that occur around the mouth and genitals

Oral candidiasis (thrush) White coating of the tongue and mouth

Candida esophagitis Painful ulceration of the esophagus

Pneumocystic carinii pneumonia (PCP) Causes fever, cough and shortness of breath

Pulmonary TB Produces cough, sputum and difficult breathing

Kaposi's sarcoma This is a cancer that usually appears as purplish-brown lesions on the skin.

Cryptosporidiosis Major symptoms are severe diarrhea and weight loss

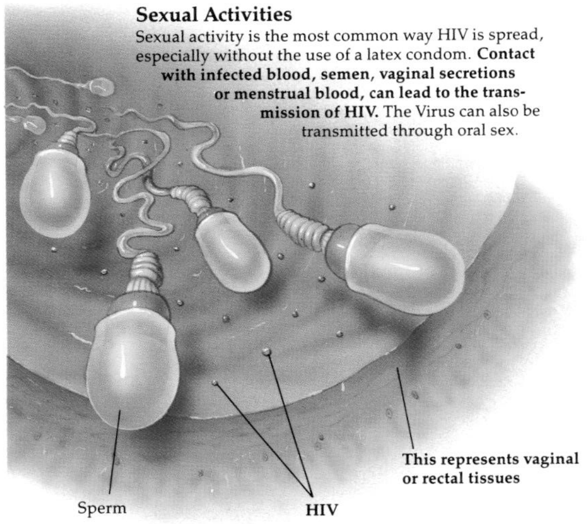

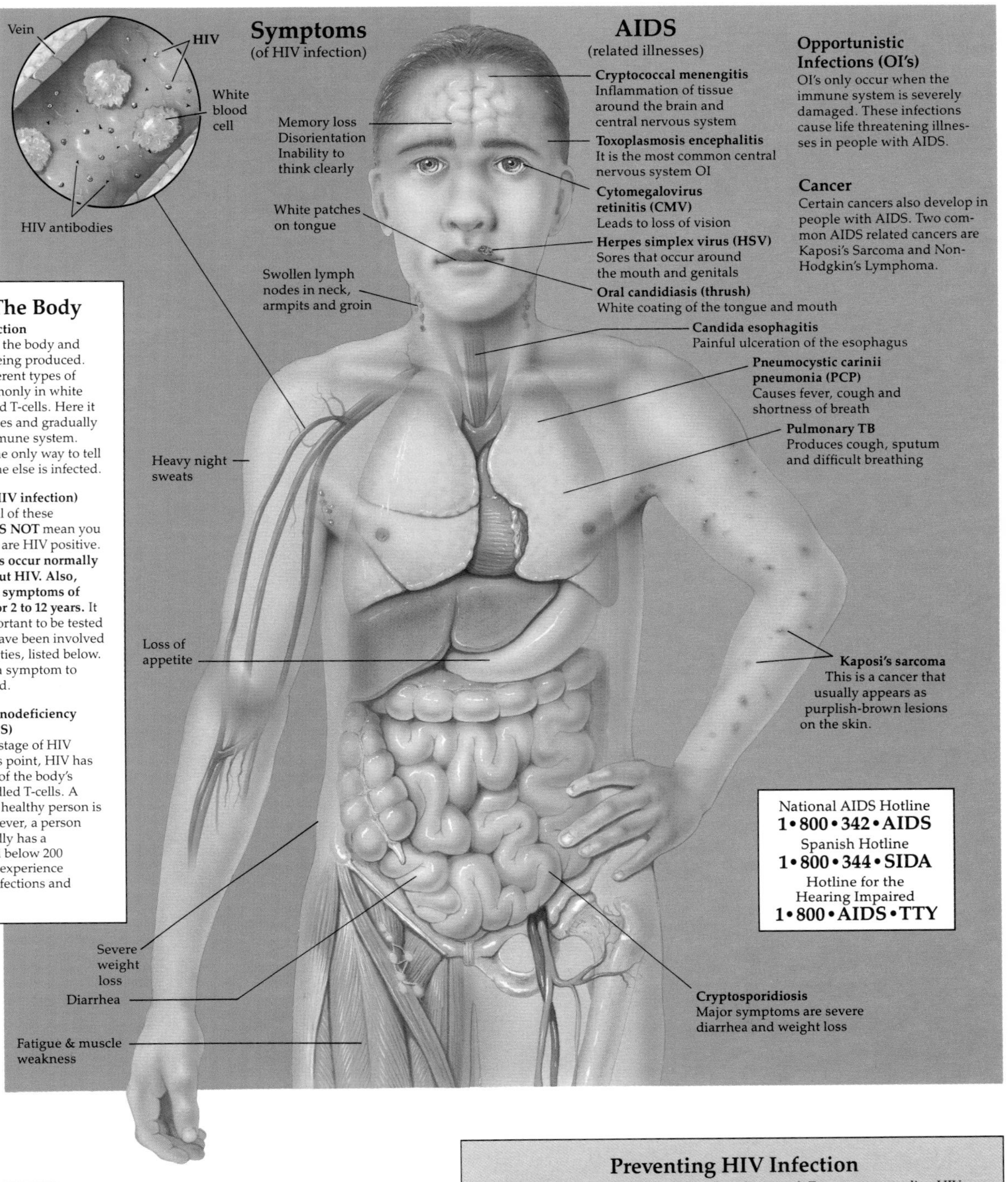

Opportunistic Infections (OI's)

OI's only occur when the immune system is severely damaged. These infections cause life threatening illnesses in people with AIDS.

Cancer

Certain cancers also develop in people with AIDS. Two common AIDS related cancers are Kaposi's Sarcoma and Non-Hodgkin's Lymphoma.

HIV and The Body

Initial HIV Infection
HIV has entered the body and antibodies are being produced. HIV lives in different types of cells, most commonly in white blood cells, called T-cells. Here it grows, reproduces and gradually weakens the immune system. An HIV test is the only way to tell if you or someone else is infected.

Symptoms (of HIV infection)
Having any or all of these symptoms **DOES NOT** mean you or someone else are HIV positive. **These symptoms occur normally in people without HIV. Also, there may be no symptoms of HIV infection for 2 to 12 years.** It is therefore important to be tested if you feel you have been involved in any risk activities, listed below. Do not wait for a symptom to occur to be tested.

Acquired Immunodeficiency Syndrome (AIDS)
AIDS is the last stage of HIV infection. At this point, HIV has destroyed most of the body's immune cells called T-cells. A T-cell count in a healthy person is about 1200; however, a person with AIDS usually has a T-cell count well below 200 or has begun to experience opportunistic infections and certain cancers.

National AIDS Hotline
1•800•342•AIDS
Spanish Hotline
1•800•344•SIDA
Hotline for the Hearing Impaired
1•800•AIDS•TTY

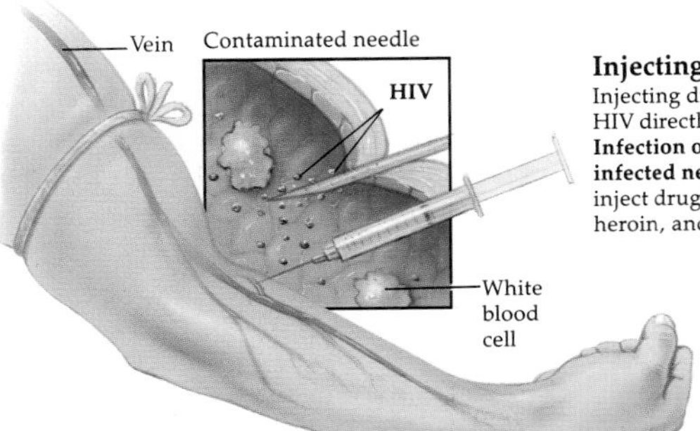

Injecting Drug Use (IDU)

Injecting drugs can transmit HIV directly into the body. **Infection occurs when sharing infected needles, and syringes** to inject drugs like steroids, heroin, and cocaine.

Vein
Contaminated needle
HIV
White blood cell

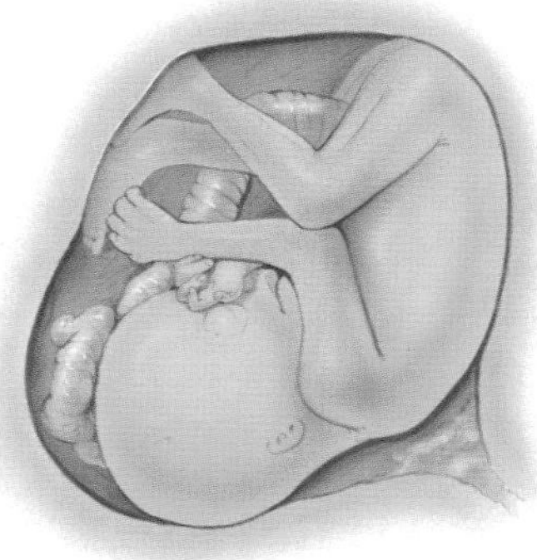

Pregnancy

Infected body fluids from the mother are passed to an infant **during pregnancy, delivery and through breast feeding.** Babies born to infected mothers have a 15% to 25% chance of becoming infected.

Treatment

Currently, AIDS is not curable. Early detection and treatment of HIV and AIDS can increase life expectancy. There are no vaccines available to date, but medications are available to slow the progression of HIV and development of AIDS. Other medications are available to protect against and treat the variety of related illnesses that may develop. Consult a doctor that specializes in care of people with HIV and AIDS for the best individual treatment.

How HIV is Not Spread

Although HIV can be found in saliva and tears, transmission is unlikely. **HIV is not spread by coughing, sneezing, crying or sweating on someone.** One cannot be infected from toilets, locker rooms, showers, towels, telephones, public areas, silverware, drinking glasses, clothing, bed linen, kissing, hugging, touching or casual contact.

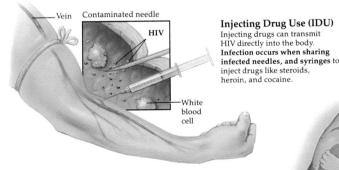

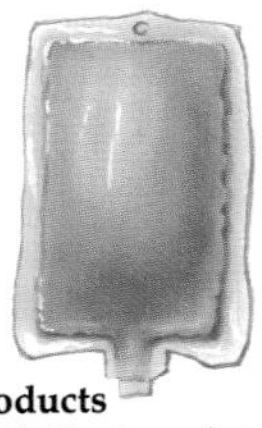

Blood and Blood Products

The highest risk of HIV infection is coming into contact with infected blood. Since 1985 the AIDS antibody test has been used to screen blood donations for the AIDS virus. Now **it is rare to become infected by receiving blood or blood products through transfusion.** People who donate blood are not at risk.

Preventing HIV Infection

The risk of catching HIV can be reduced or eliminated. To prevent spreading HIV, use the following methods:

SEX: The best protection is not to have sex until after having established a long-term mutually monogamous relationship with an uninfected person. Ask for an AIDS test before sexual relationships. Use a latex condom with spermacide during sex. **Latex condoms do not provide 100% protection, but have been proven to reduce the risk of HIV infection.**

Health: Stay healthy and free of Sexually Transmitted Diseases (STD's). **STD's can create open sores** and inflammation on the skin, around and inside the mouth, genital and anal areas **which provide a direct route of entry for HIV.**

Drugs: Not using drugs or having sex with **injecting drug users reduces the risk of HIV infection.** If injecting drugs, never share needles. Obtain sterile needles from a pharmacy or through a needle exchange program. To decontaminate needles and syringes, flush them 3 or 4 times with chlorine bleach and repeat with tap water. This reduces, but does not eliminate the risk of HIV infection.

Pregnancy: An HIV test should be taken before becoming pregnant if you have engaged in high risk behavior. The best protection for the infant is if the mother has tested HIV negative before, during, and after pregnancy and breast feeding.

Risks for Becoming Infected with HIV

HIGH RISK	MEDIUM RISK	LOW RISK
• Anal, vaginal or oral sex (no condom)	• Anal, vaginal or oral sex (with condom)	• Intimate touching
• Sharing needles and syringes (not cleaned with chlorine bleach)	• Sharing needles and syringes (cleaned with chlorine bleach)	• Blood transfusion (after 1985)
• Sex with injecting drug users	• Infected body fluids on broken skin	• Organ transplant
• Multiple sex partners		• Deep kissing

UNDERSTANDING MIDDLE EAR INFECTIONS

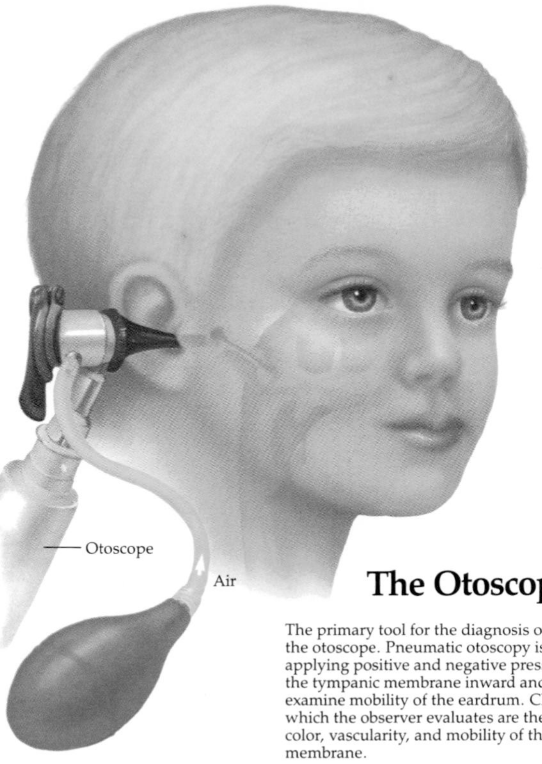

Otitis Media

Otitis Media is an inflammation in the middle ear. Among the causes are bacteria, viruses or allergies and malfunction of the eustachian tube. There may be an accumulation of fluid in the middle ear. This fluid may restrict the movement of the tympanic membrane and middle ear structures, resulting in a decrease in hearing. Pain may occur if there is pressure on the eardrum and fever may result from the products of infection.

Pneumatic System

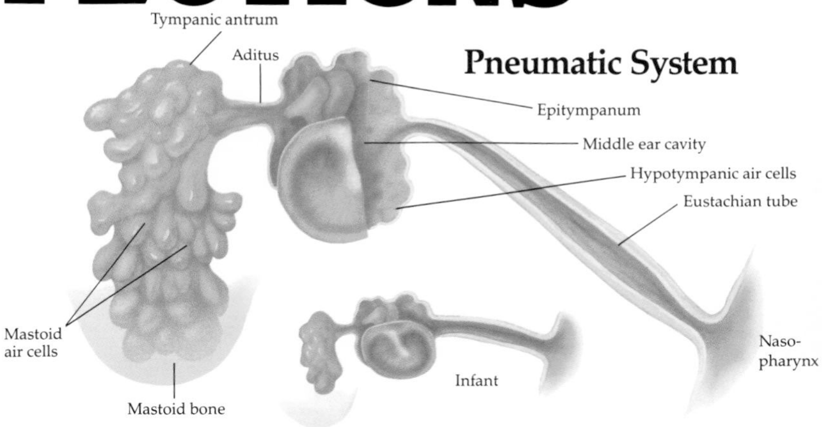

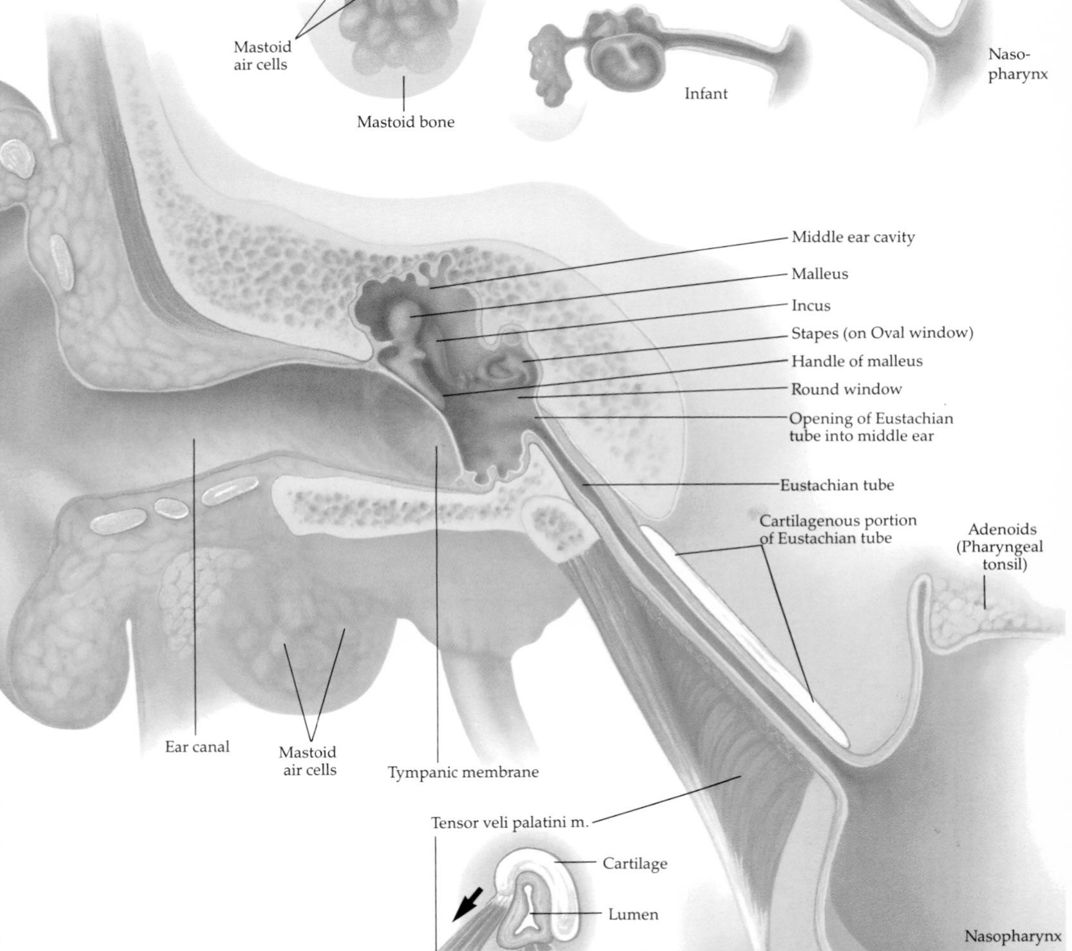

The Otoscope

The primary tool for the diagnosis of ear disease is the otoscope. Pneumatic otoscopy is performed by applying positive and negative pressure to move the tympanic membrane inward and outward to examine mobility of the eardrum. Characteristics which the observer evaluates are the position, color, vascularity, and mobility of the tympanic membrane.

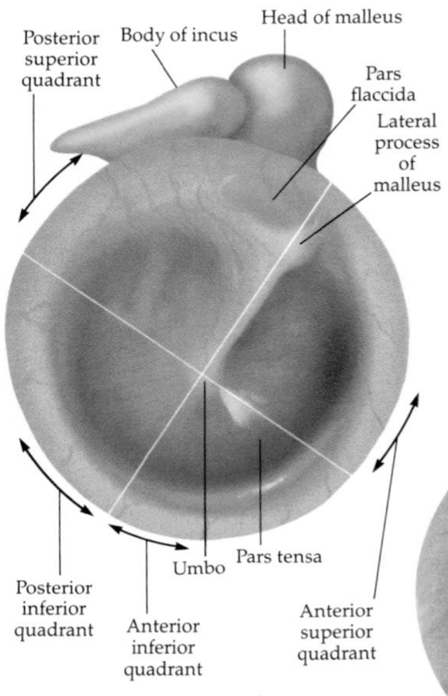

Tympanic Membrane

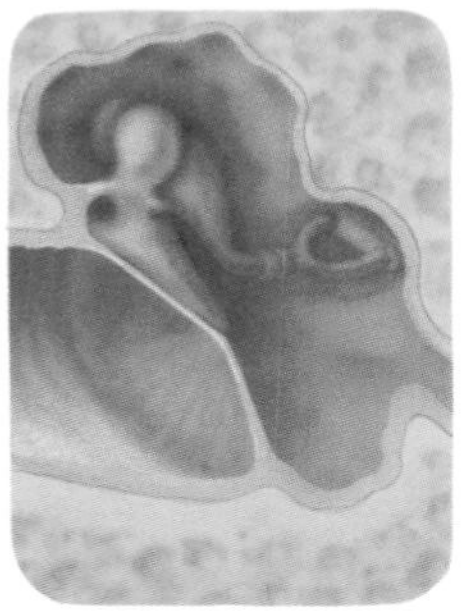

Normal

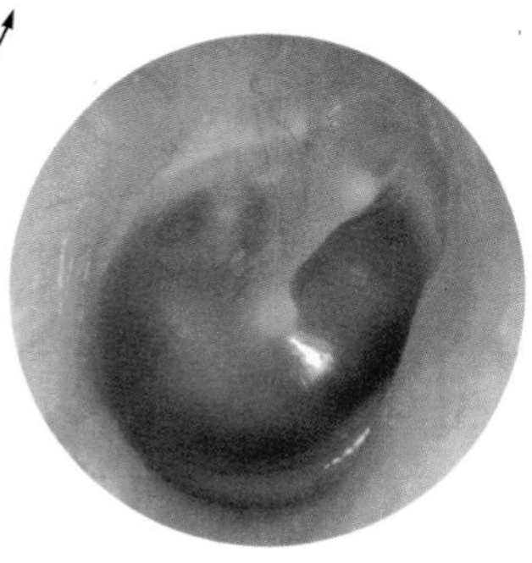

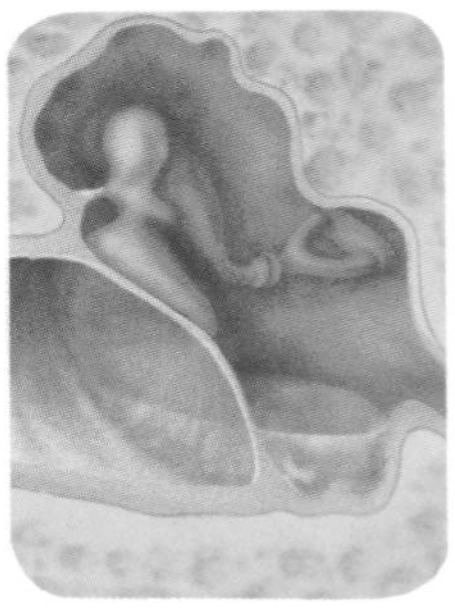

Otitis Media with Effusion

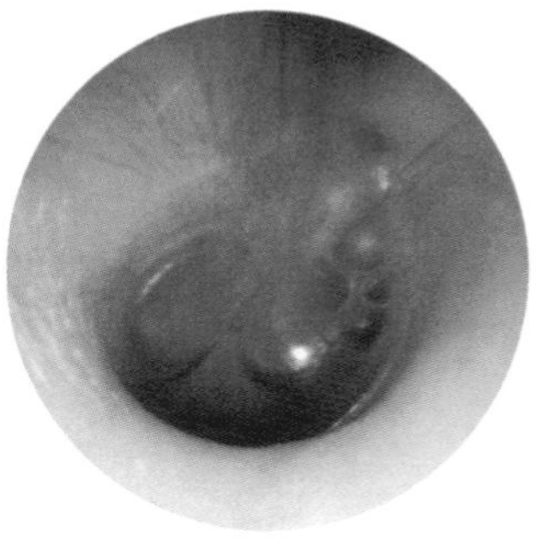

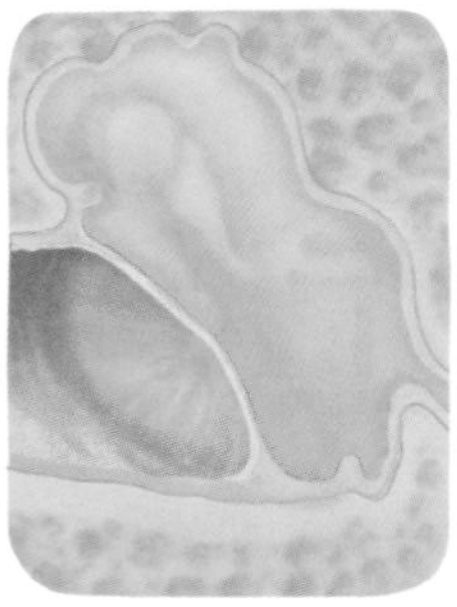

Secretory Otitis Media

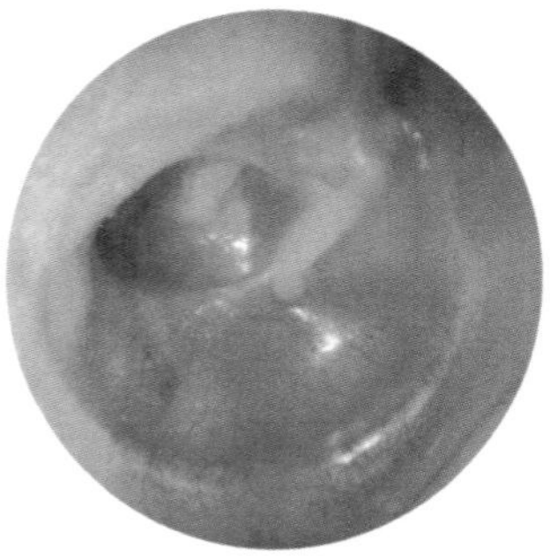

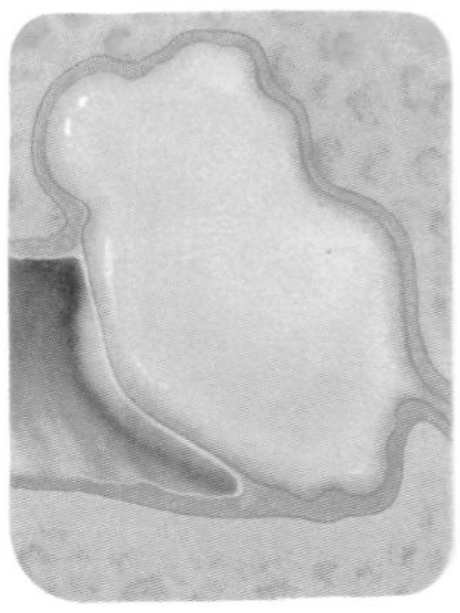

Acute Otitis Media

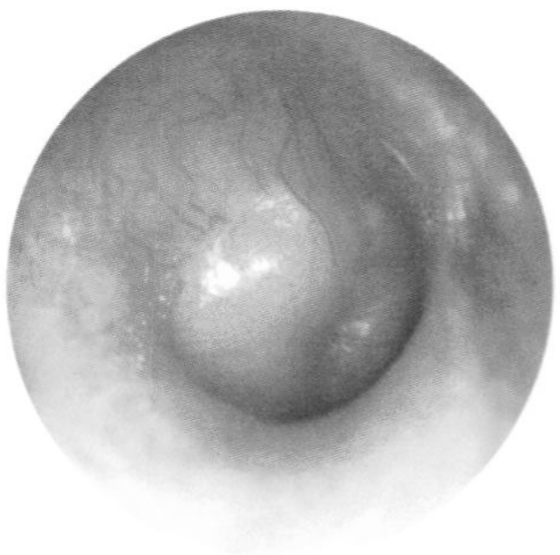

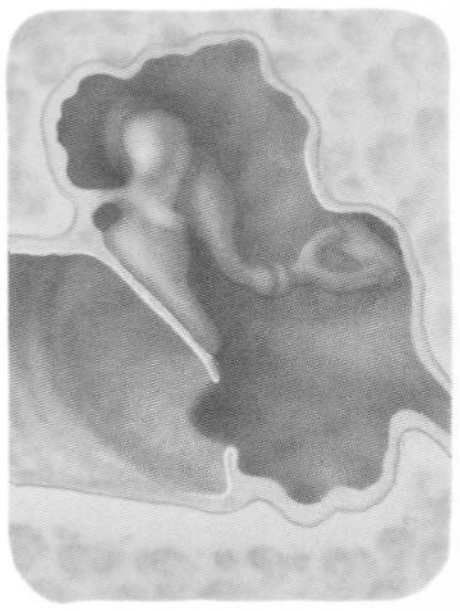

Perforation

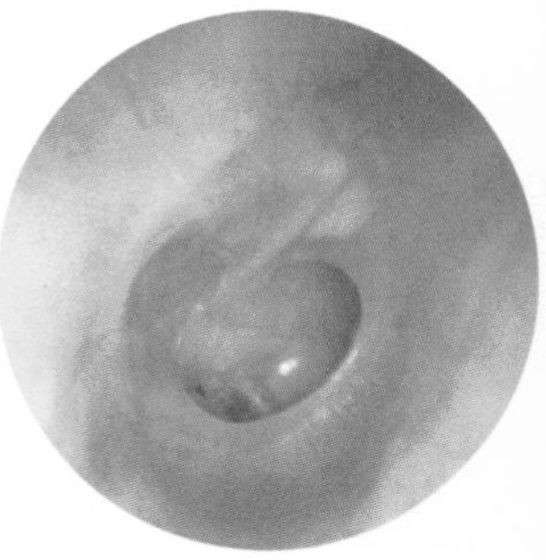

Tympanometry

(Configuration varies with equipment)

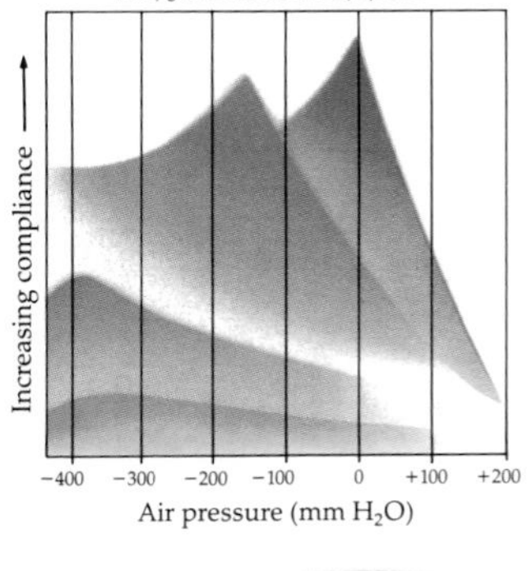

Audiogram

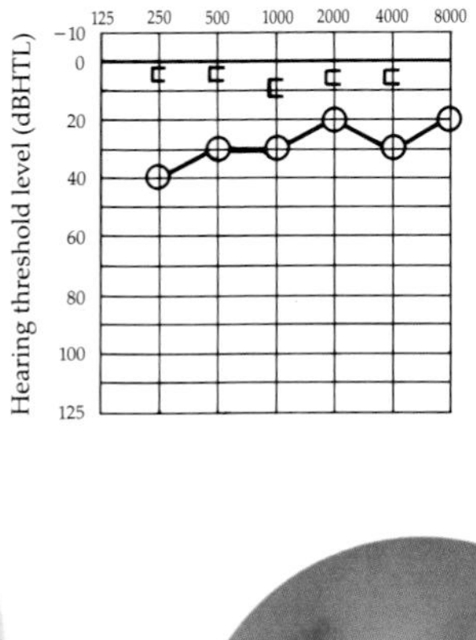

Typical hearing loss caused by fluid in the middle ear

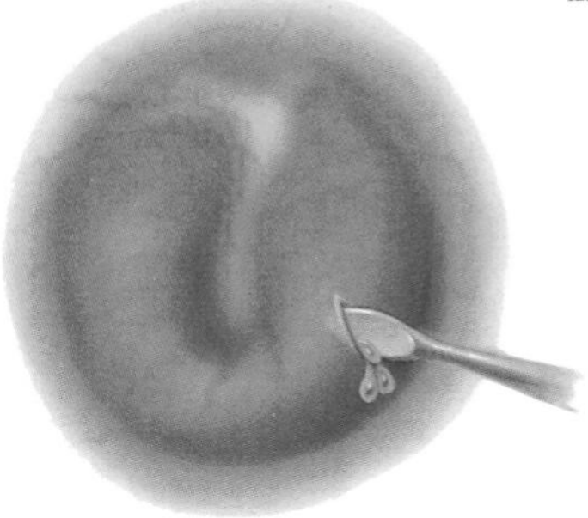

Myringotomy

Myringotomy is an incision through the tympanic membrane.

Ventilation Tube

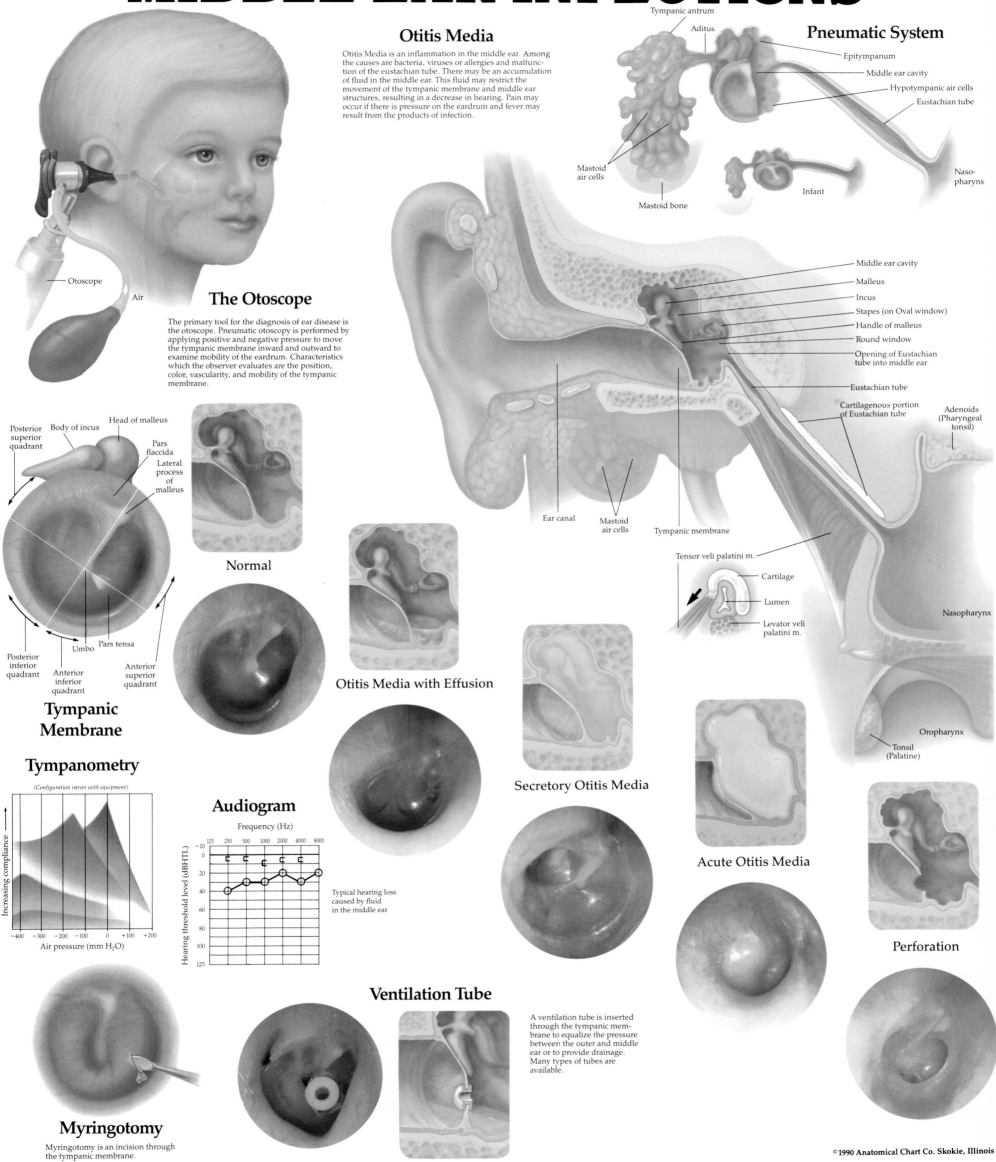

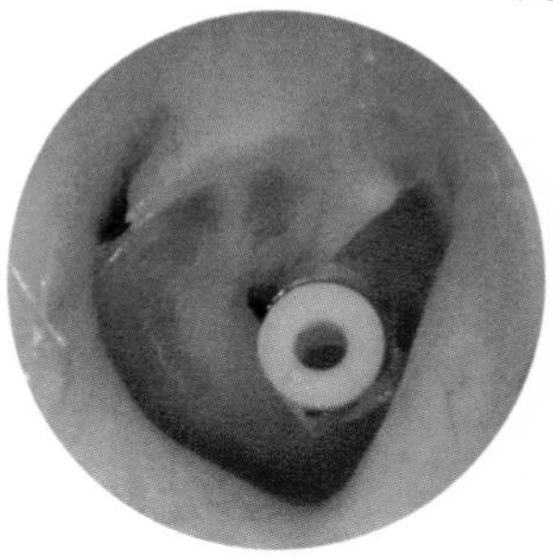

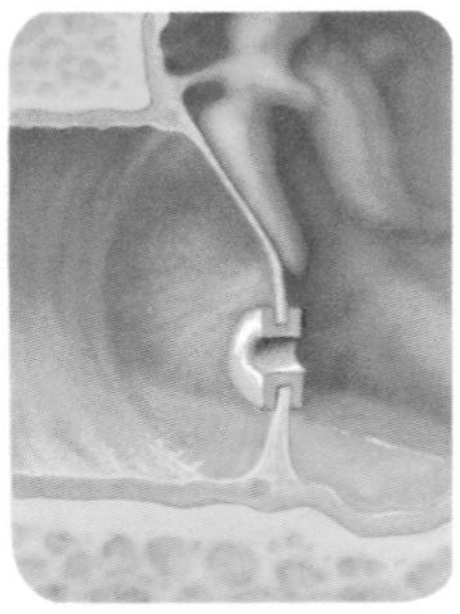

A ventilation tube is inserted through the tympanic membrane to equalize the pressure between the outer and middle ear or to provide drainage. Many types of tubes are available.

UNDERSTANDING THE COMMON COLD

What Is a Cold?

The common cold or upper respiratory infection is a contagious inflammation of the mucous membranes of the head and throat. It is characterized by a runny nose (clear discharge), sneezing, dry cough, and sore throat and is often accompanied by headaches, muscle aches, and mild fever. Most colds are caused by viruses and usually clear up after a few days with symptomatic treatment. More serious bacterial infections of the ears, sinuses, throat, and lungs may follow a viral cold.

Respiratory Mucosa

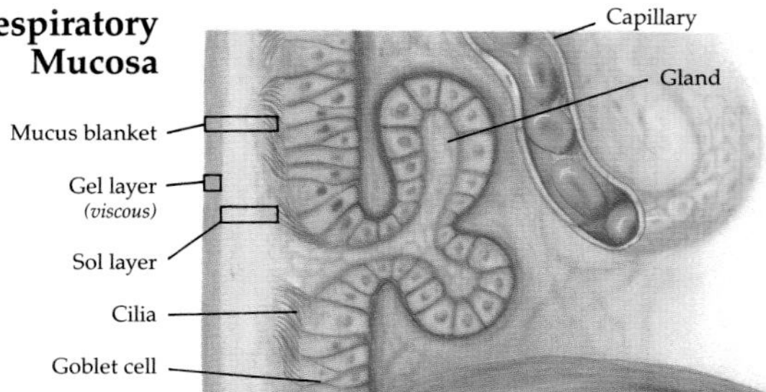

What Is Inflammation?

With massive invasion of foreign microorganisms, such as a virus, our body responds with a full defense called *inflammation.* This response is characterized by redness, heat, swelling, and pain. Increased circulation causes redness in the affected tissues. Delivery of more blood and heat from the interior of the body, known as fever, helps the body's defenders to function more effectively and slows the growth of the invading organism. Fluid from blood vessels leaks into an inflamed area, causing swelling and delivering antibodies. Identified invaders are then broken down and devoured by granulocytes. During a cold, the resulting debris or exudate is blown from the nose or coughed up from the air passages.

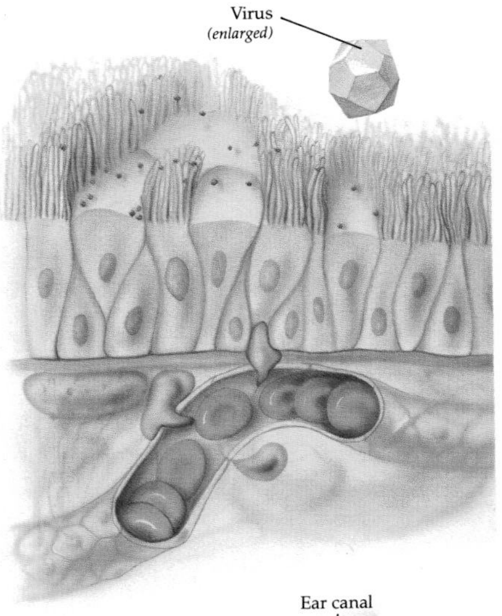

The Nose

As we breathe, foreign particles, viruses, and bacteria adhere to the mucus covered walls of the nose. The mucous membranes of the adjacent sinuses produce a clear mucus that washes the inside of the nose. *Acute rhinitis* is the most common of the upper respiratory infections and is one of the first discomforts of the common cold. It is characterized by swelling and an increased discharge from the mucous membranes lining the nose–a *runny nose.*

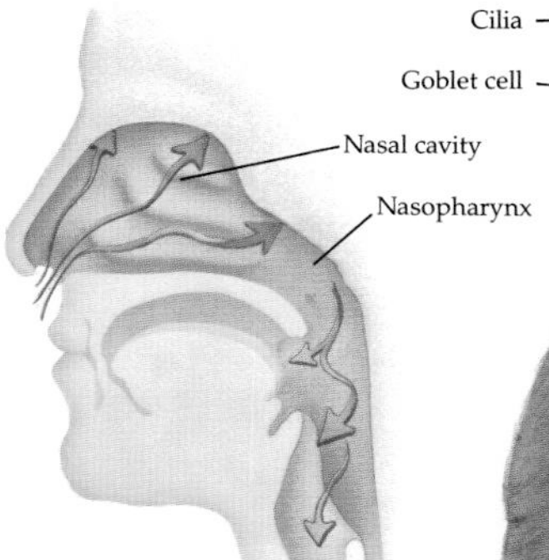

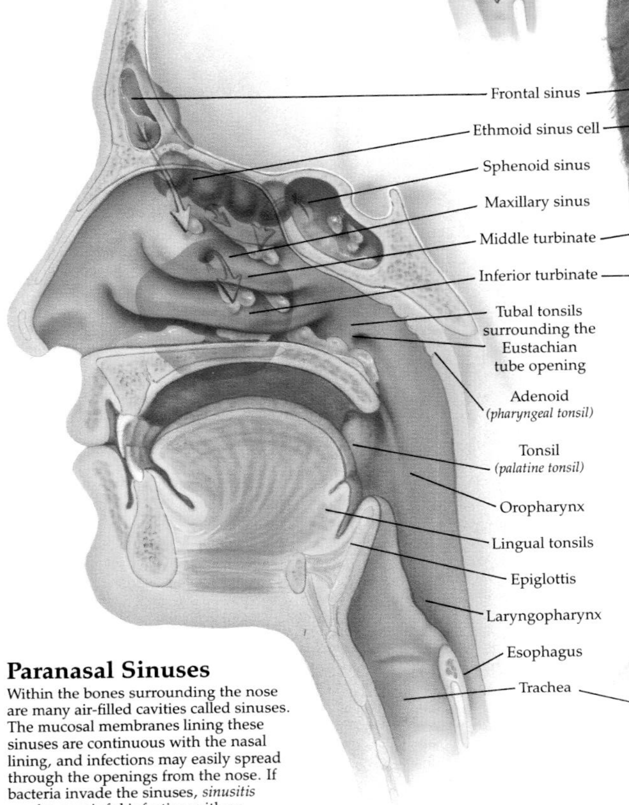

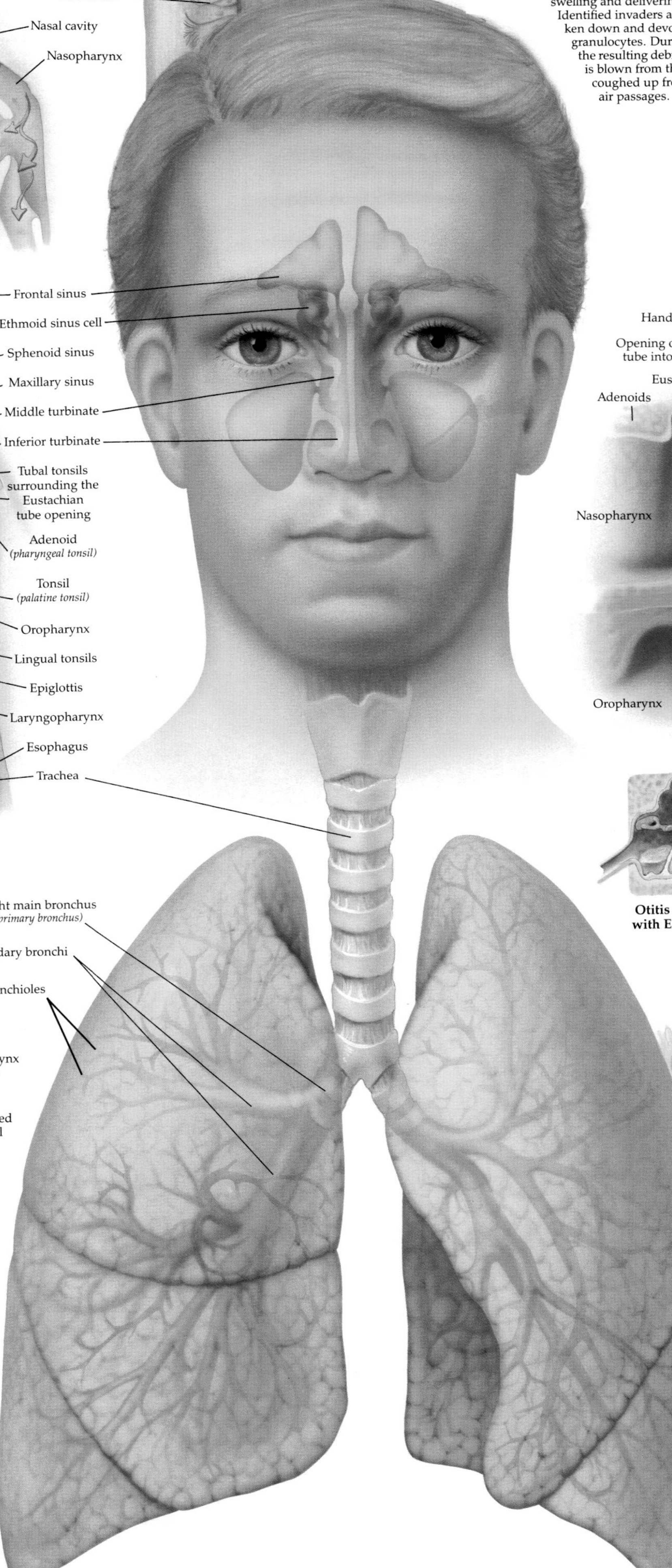

Paranasal Sinuses

Within the bones surrounding the nose are many air-filled cavities called sinuses. The mucosal membranes lining these sinuses are continuous with the nasal lining, and infections may easily spread through the openings from the nose. If bacteria invade the sinuses, *sinusitis* results, a painful infection with an associated gray, yellow, or green discharge from the nose.

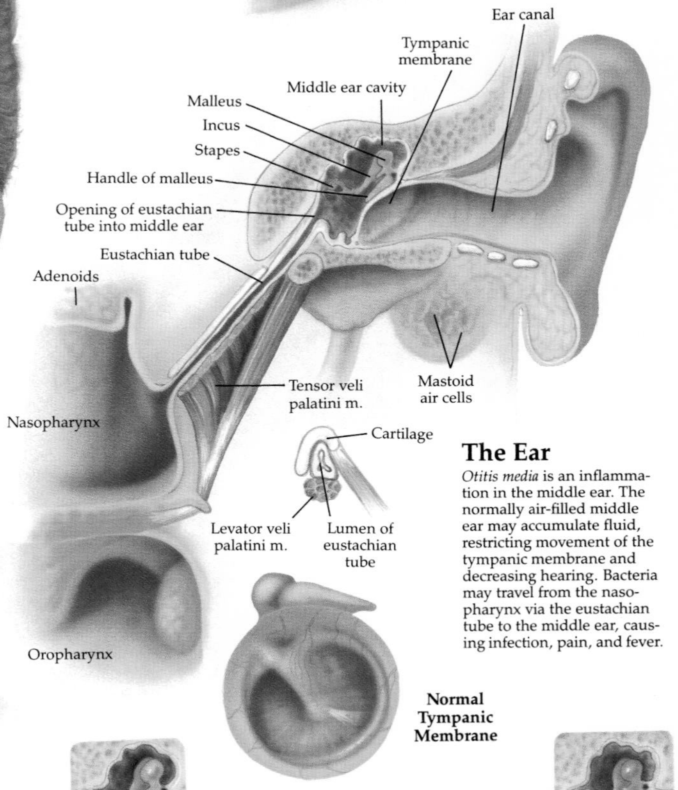

The Ear

Otitis media is an inflammation in the middle ear. The normally air-filled middle ear may accumulate fluid, restricting movement of the tympanic membrane and decreasing hearing. Bacteria may travel from the nasopharynx via the eustachian tube to the middle ear, causing infection, pain, and fever.

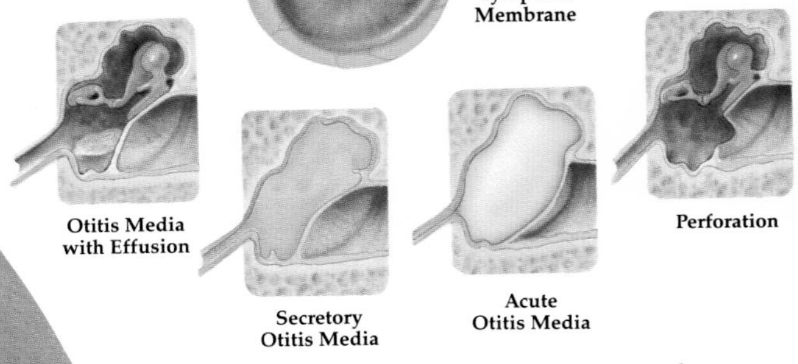

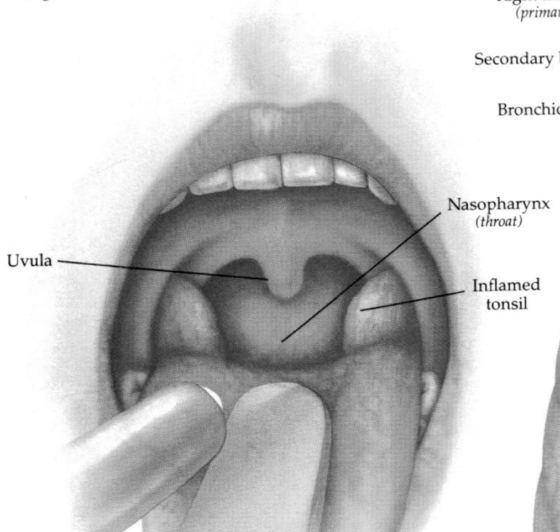

The Throat and Tonsils

Airborne particles, viruses, and bacteria not trapped in the nose or from the mouth may land on the pharynx or back of the throat. Here, a circle of lymphatic tissue (Waldeyer's ring) surrounds the pharynx and serves as a first line of defense. This lymphatic ring is formed of pharyngeal tonsils, tubal tonsils, palatine tonsils, and lingual tonsils. Inflammation affecting the throat and tonsils causes swelling, redness, and pain when swallowing and is commonly called a *sore throat.* Bacteria (most frequently *Streptococcus*) may cause a more severe infection called *strep throat.* An infection of the palatine tonsils is called *tonsilitis,* and their enlargement may cause discomfort. The lymph nodes of the neck may also become enlarged as they respond to fighting the infection of a cold.

The Trachea and Bronchi

An acute inflammation of the tracheobronchial tree, called *bronchitis,* often develops after a common cold. Increased blood flow and swelling of the mucous membranes disrupts the protective membrane and bronchial cilia, allowing bacterial invasion. Coughing effectively expels the inflammatory products: mucus, bacteria, and lumps of phlegm accumulating in these respiratory passages.

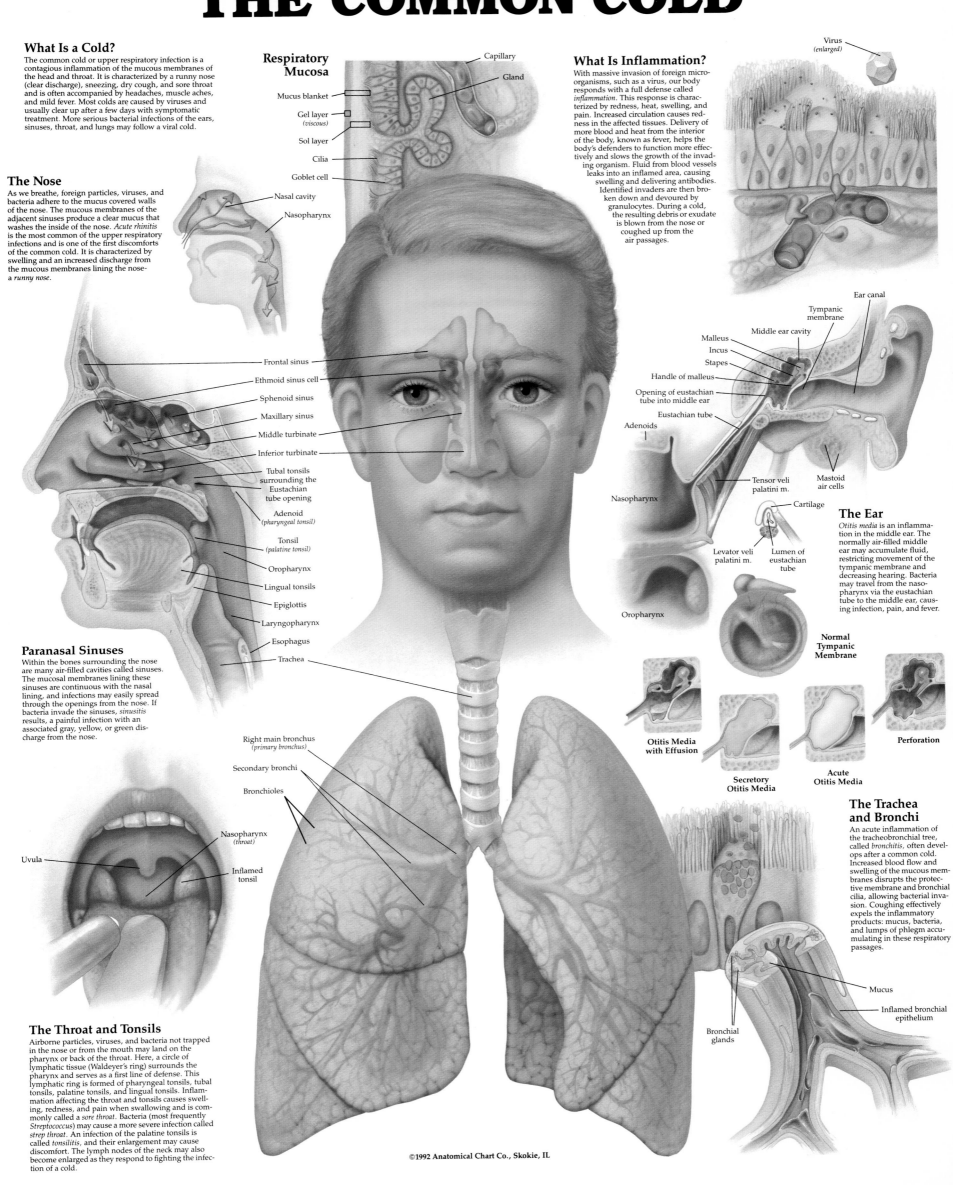

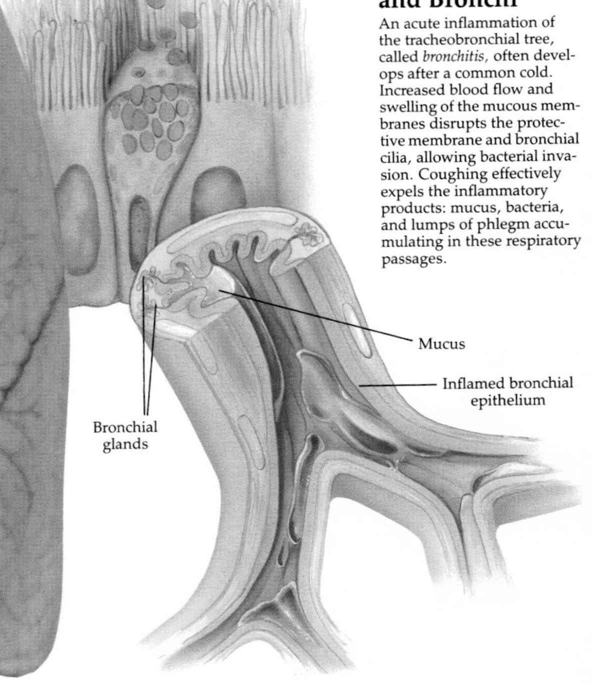